Lecture Notes in Economics and Mathematical Systems   543

Founding Editors:

M. Beckmann
H. P. Künzi

Managing Editors:

Prof. Dr. G. Fandel
Fachbereich Wirtschaftswissenschaften
Fernuniversität Hagen
Feithstr. 140/AVZ II, 58084 Hagen, Germany

Prof. Dr. W. Trockel
Institut für Mathematische Wirtschaftsforschung (IMW)
Universität Bielefeld
Universitätsstr. 25, 33615 Bielefeld, Germany

Editorial Board:

A. Basile, A. Drexl, H. Dawid, K. Inderfurth, W. Kürsten, U. Schittko

Brüggen
Celanos

Christoph Benkert

Default Risk in Bond and Credit Derivatives Markets

Author

Christoph Benkert
Kreuznacher Straße 44
60486 Frankfurt/Main
Germany

Library of Congress Control Number: 2004106908

Chapter 5 is a preprint of an article published in the Journal of Futures Markets, Vol. 24 No. 1 (John Wiley & Sons, NJ, 2004)

ISSN 0075-8442
ISBN 3-540-22041-0 Springer Berlin Heidelberg New York

This work is subject to copyright. All rights are reserved, whether the whole or part of the material is concerned, specifically the rights of translation, reprinting, re-use of illustrations, recitation, broadcasting, reproduction on microfilms or in any other way, and storage in data banks. Duplication of this publication or parts thereof is permitted only under the provisions of the German Copyright Law of September 9, 1965, in its current version, and permission for use must always be obtained from Springer-Verlag. Violations are liable for prosecution under the German Copyright Law.

Springer is a part of Springer Science+Business Media

springeronline.com

© Springer-Verlag Berlin Heidelberg 2004
Printed in Germany

The use of general descriptive names, registered names, trademarks, etc. in this publication does not imply, even in the absence of a specific statement, that such names are exempt from the relevant protective laws and regulations and therefore free for general use.

Typesetting: Camera ready by author
Cover design: *Erich Kirchner*, Heidelberg

Printed on acid-free paper   42/3130Di   5 4 3 2 1 0

Preface

This work has been accepted as my doctoral thesis by the Faculty of Business Administration and Economics at Johann Wolfgang Goethe-University in Frankfurt. In writing this book on credit risks, I have accumulated substantial debt on my personal balance sheet. There are a number of people whom I have to thank for bearing this kind of credit risk. Thus, here follows my personal list of IOUs.

First of all, I am indebted to Prof. Christian Schlag. His exceptional commitment to doctoral education at his Chair for Derivatives and Financial Engineering has created the supportive environment that was in every respect ideal for my research. I would like to thank him most sincerely for his academic and personal interest in my work.

I would also like to thank Prof. Mark Wahrenburg for many constructive comments and for reviewing my thesis together with Profs. Uwe Hassler and Dietrich Ohse.

I am very grateful to my colleagues, Nicole Branger, Angelika Esser, and Burkart Mönch for the cooperative atmosphere in our group and their help with various tricky questions. Burkart's patient IT and logistic support warrants special thanks.

My thanks go also to Michael Belledin with whom it always has been a great pleasure to make use of various hill climbing routines in trying to reach local and global extrema – in front of a computer's screen almost as much as on the bike.

I have found many friends in the University of Frankfurt's graduate program in Finance and Monetary Economics and I would like to thank Alexander, Sir Toby, O_2, Mr. PC, Keith, and Iskra for the many interesting discussions we had when we did not just have fun. I owe special thanks to Micong Klimes for her invaluable help, comments, and extremely generous credit terms.

Finally, I wish to thank all of my family, who have supported me in every conceivable way. My parents have been my lenders of last resort, both spiritually and materially. Tony has provided huge leverage to my life. And as there can't be any credit without some equity held against it – thanks, Annette, for trusting me enough to hold on to your stake.

Frankfurt, May 2004 *Christoph Benkert*

Contents

1 Introduction 1

2 On the Economic Content of Models of Default Risk . 7
   2.1 Introduction 7
   2.2 A Criterion for Economic Interpretability 8
   2.3 Models of Default Risk 9
      2.3.1 Reduced-Form Models 9
      2.3.2 Firm Value Models 11
      2.3.3 Hybrid Approaches 13
   2.4 Interpretability of Firm Value Models 14
   2.5 Conclusion 20

3 Intensity-Based Modeling of Default 21
   3.1 Introduction 21
   3.2 Default Arrival and the Default Event 22
   3.3 The Hazard Rate 24
   3.4 Loss Given Default 25
      3.4.1 Nature of the Recovery Process 25
      3.4.2 Recovery Regime 26
   3.5 Defaultable Bond Prices 27
   3.6 Implications for the Empirical Studies 30
   3.7 Affine Term Structure Models in the Context of Default Risk 31
      3.7.1 Model Description 32

         3.7.2 Completely Affine Models with Independent
               Factors 33
         3.7.3 Incorporating Correlation between Risk-Free and
               Risky Rates 36
         3.7.4 Maximal Models: Essentially Affine Specifications. 39
    3.8 Summary and Outlook 41

4 **The Empirical Performance of Reduced-Form
   Models of Default Risk** 43
    4.1 Preliminaries 43
         4.1.1 Data Description 43
         4.1.2 Defaultable Term Structure Estimation 44
         4.1.3 Risk-Free Term Structure Estimation 49
         4.1.4 Discussion of Data Quality 50
    4.2 Estimation of Completely Affine Term Structure
        Models for Defaultable Rates 53
         4.2.1 Estimation Technique 54
               4.2.1.1 State-Space Representation 54
               4.2.1.2 State-Space Specification 55
               4.2.1.3 Kalman Filter Mechanism 57
         4.2.2 Implementation 59
         4.2.3 Results 60
               4.2.3.1 Preferred Models 60
               4.2.3.2 In- and Out-of-Sample Fit 63
               4.2.3.3 Parameter Estimates 65
    4.3 Estimation of Completely Affine Term Structure
        Models for Spreads 69
         4.3.1 Implementation 70
         4.3.2 Results 70
               4.3.2.1 Preferred Models 70
               4.3.2.2 In- and Out-of-Sample Fit 72
               4.3.2.3 Parameter Estimates 72
    4.4 Incorporating Correlation 73
         4.4.1 Implementation 73
         4.4.2 Results 74

| | | 4.4.2.1 In- and Out-of-Sample Fit .... 75 |
| | | 4.4.2.2 Parameter Estimates .... 77 |

4.5 Estimation of Essentially Affine Term Structure Models for Defaultable Rates 79
    4.5.1 Estimation Technique: Efficient Method of Moments 79
    4.5.2 Implementation 83
    4.5.3 Results 84
        4.5.3.1 Auxiliary Model 84
        4.5.3.2 Structural Model 86

4.6 Summary 89

5 Explaining Credit Default Swap Premia 91
  5.1 Introduction 91
  5.2 Modeling Idea 94
  5.3 Data 98
  5.4 Estimation and Results 103
  5.5 Robustness Checks 107
  5.6 Conclusion 109

6 Conclusion 111

Appendix 115

A Calculation of Volatility Proxies 115

B Tables for Chapter 4 117

C Tables for Chapter 5 123

References 131

1

Introduction

Loans are probably the oldest financial contracts. Given the long history of money lending, it is astounding that the assessment and management of credit risk is still one of the most urgent issues today. In fact, the topic has experienced an unpleasant comeback as default rates have surged recently. From 2000 to 2001, the number of defaulted firms included in the Merrill Lynch Global Bond Index has doubled to 76. The ratio of three rating downgrades for every upgrade in 2001 provides evidence on the deterioration of overall credit quality. As a consequence of the increase in uncertainty, spreads on publicly traded issues have fluctuated widely.

On the other hand, the outstanding volume of credit contracts and hence their economic importance are inconceivably large. The size of the world bond market excluding emerging markets has grown from 13 trillion USD in 1990 to 33 trillion USD in 2001. Therefore, even in the absence of spectacular credit events, such as the default of the investment grade rated debt of Enron, a constant interest to improve credit risk management is warranted. The introduction and use of innovative financial instruments, such as credit derivatives and collateralized debt obligations, have further increased the need for ever better credit risk models.

The importance of the topic for the financial system is reflected in the regulatory work of the Basel committee on Banking Supervision. The attention that the new Basel Capital Accord on which the third consultative paper has just been issued has received is unparalleled. Many of the requirements that the Basel committee plans to impose on creditors have already influenced the design of models of default risk. One of the most prominent examples is the stipulation that the probability of a loss on a loan and the severity of this loss should be modeled separately.

The focus of this book is on the empirical investigation of default risk as it becomes manifest in credit default swap premia and the spreads of corporate bonds over government bonds. Both of these data sources are examined separately, and the contributions of the respective studies to the existing research are the following. For the first time, firm-specific term structures are used in the estimations and both features, exploring default risk on the level of firms as well as using term structures, deserve notice. The data used in existing empirical work on default risk – which is rare in the first place – usually consist of prices of individual bonds and are often pooled by ratings. For the estimation of term structure models that we will pursue, information on the whole term structure is more appropriate. We test a number of competing models both on defaultable rates and spreads and are the first to be able to compare the results across two different currencies, namely USD and EUR. In a clinical study, we explore essentially affine models, which extend the class of completely affine models and have not been considered in a default risky context up to now.

Hardly any empirical research has been conducted using credit default swap data. We enter this new ground by explaining credit default swap premia using rating and accounting information as well as equity volatility data. An innovative feature is the inclusion of historical and implied volatility on a firm level, rather than an index level.

The emphasis on empirical work reflects the fact that the theoretical research on default risk is already quite advanced and extensive. Hence, we find it more timely to address the question of how successful the proposed models are in describing the available data. However, as theoretical research in the field matures, factoids tend to emerge. One of these is the popular notion that *all* models in the class of firm value models are associated with the feature of economic interpretability. We make a theoretical contribution by revisiting the so-called structural approach with the purpose of illuminating this (mis)conception.

An important assumption that will be made throughout all chapters – and that is certainly debatable – is that credit default swap premia and corporate bond spreads are indeed entirely due to credit risk. This means that we abstract from liquidity risk which may be present in an investment in such securities. We also make no attempt to quantify the impact of regulatory decrees or taxes.

Another field of research that we refrain from including in our analyses is that of strategic defaults. The research questions in this area, the modeling environment in which they are addressed, and the techniques used are so different that it would be far beyond the scope of this study to consider this issue simultaneously.

The decision not to consider other factors that potentially influence credit spreads has been taken for two reasons. On a theoretical ground, most of these factors are yet to be modeled convincingly. Take, for example, liquidity risk. It is far from clear, which of the competing measures should be used. There is even less evidence on the time series behavior of this factor. And there is not even an agreement on how investors are affected by the presence of this risk, i.e. how prices are quantitatively affected.

From a practical point of view, a treatment of these issues would complicate the models to such an extent that their estimation – given the available data – would become infeasible. An exception to this decision is the attempt to control for liquidity effects in the examination of credit default swap premia in chapter 5. However, the liquidity proxy that is included in this study is not informative. But before we get ahead of ourselves, we provide a guide to the chapters of this book in the following.

In chapter 2, we briefly review the most prominent approaches that have been developed in the academic literature to model default risk. Since the empirical studies are performed in several so-called reduced-form settings, particular attention is devoted to the question whether this decision is appropriate in economic terms. It is frequently put forward that structural models of default – i.e. those that explicitly model the process of a firm's value – offer economic interpretability. From an academic point of view, this would be a favorable feature. Unfortunately, by construction, the default event in reduced-form models cannot be interpreted economically. Nevertheless, there is a justification of our choice of model with respect to this aspect: We argue that formulations of firm value models that are capable of reproducing even the simplest empirical facts make assumptions (sometimes implicitly) that severely interfere with economic intuition.

Chapter 3 lays out the model context which is primarily used in the estimations. Strictly concentrating on the gist of the applicable literature, the concept behind the intensity-based modeling of default is presented. We address the notion of the default event in this literature and the two basic elements of this approach, the hazard rate and the recovery assumption. Then we show how defaultable bond prices can be derived in this setup. The chapter concludes with a treatment of affine term structure models in the context of default risk. This model class is a popular choice in the literature of the risk-free term structure. We estimate various specifications using defaultable bond price data.

The empirical work is presented in chapters 4 and 5. We discuss the bond data, respective estimation methodologies, and results in chapter 4. In this chapter, we rely heavily on the preparatory work of chapter 3. In the starter case, the defaultable rates and spreads, respectively, are driven by independent Vasicek and Cox-Ingersoll-Ross processes. In a second step, we allow for correlation between the risk-free interest rates and spreads following Duffee (1999) [25]. The estimation of these models is carried out by quasi-maximum likelihood via a Kalman filter recursion. In a clinical study, we explore a more flexible form of the market price of risk due to Duffee (2002) [26]. For estimation, we use the efficient method of moments methodology proposed by Gallant and Tauchen (1996) [38].

Chapter 5 is essentially self-contained. It uses data on credit default swaps and analyzes these data using a much simpler model. Estimation is carried out in a reduced-form context as well, but the regression model that is employed offers a connection to firm value models. Specifically, we investigate the effects of historical and option implied equity volatility on credit default swap premia, thus extending an idea proposed by Campbell and Taksler (2002) [17] in the context of corporate bond yields.

We conclude in chapter 6 with a summary of the results and a discussions of their implications for further research as well as policy making.

2

On the Economic Content of Models of Default Risk

2.1 Introduction

The existing theoretical literature on default risk modeling has pursued two different approaches to model default risk, firm value models (FVM) and reduced-form models (RFM). The starting point of the first approach is a stochastic differential equation for the firm value. Default occurs when the value of the firm hits a certain boundary, which is typically specified in terms of the value of some liabilities. Thus, the firm value process along with the current debt ratio determines the probability of default. This idea has been laid out in the seminal papers by Black and Scholes (1973) [12] and Merton (1974) [53]. Since then, it has been employed frequently, not least due to the fact that it appeals to our natural intuition of the economic meaning of bankruptcy. In contrast, RFM do not model the economics underlying default risk in this sense but purposely restrict the perception of default events to their statistical properties.

In the simple setup of the Merton model, the cause of default is so perspicuous that today it is an established truth and a praised advantage of FVM, that within their realm the event of default is economically meaningful, while in RFM it is not. This notion is undoubtedly appropriate as far as RFM are concerned. However, with respect to FVM, it holds only for the Merton model. In more recent contributions to the literature on FVM that will be considered in this chapter, many of the restrictive assumptions of the original Merton model have been relaxed. While this increased the applicability of the models and improved their performance, it came at the cost of giving up much of their economic interpretability. Many researchers still tend to generously overlook this fact.

Since the focus of this dissertation clearly is to assess the empirical performance of reduced-form credit risk models, it is – at least from an academic's perspective – worthwhile to begin with a discussion of the economic interpretability of the respective models. This will give us the chance to review the setups of the different models. We will argue that the difference in terms of economic intuition between the two classes of models is much smaller than one would expect based on the prevailing perception.

2.2 A Criterion for Economic Interpretability

Let us begin by formulating the criterion which we will use to assess economic interpretability. It is rather strict and requires that in a given model all relevant components of default risk are considered endogenously and economically meaningful. The relevant components of default risk are (1) the probability that a borrower goes bankrupt and (2) the severity of the loss that the lenders suffer in this case. By economically meaningful we mean that the default event must be triggered by an economic mechanism as opposed to judicial decisions or the like. One can safely think of the trigger as the situation in which the value of the assets of the borrower is insufficient to cover the liabilities. This points to FVM once again, and the discussion of the Merton model in section 2.3.2 will indeed be helpful in understanding this criterion completely.

Sometimes the deterioration of credit quality is mentioned as a component of credit risk. This is of course a sensible thing to look at if one is interested in the assessment of value at risk measures over the life of a bond. It is not a component of default risk in the sense introduced above, though. Rather, it focuses on the development of the two components over time. When a term structure exists for both the probability of bankruptcy and the loss given default, it is also possible to derive probabilities for arbitrary degrees of credit deterioration before maturity.

Note that even more restrictive criteria could be applied by requiring a model to endogenize the bargaining process that inescapably follows a default or other issues deeply rooted in the field of corporate finance. In line with the research agenda laid out in chapter 1 we ignore such issues in this book. This is appropriate for the FVM we consider, as they are all set up in a perfect market. Keeping the conceptual decomposition of default risk and the criterion built on it in mind, we now turn to the alternative default models.

2.3 Models of Default Risk

2.3.1 Reduced-Form Models

It may seem surprising for a survey of models of default risk to cover reduced-form models first. After all, RFM have been developed only after FVM had already been around for about 20 years. And, admittedly, the economics of the Merton model provide for a good starting point. The reasons for the arrangement chosen in this chapter are two-fold. First, RFM will be covered in more detail in the following chapters. Second, the order is owed to the line of argumentation, which – although it is by no means meant to be – may seem somewhat destructive. In making the case for RFM, we do not attempt to claim the existence of economic intuition where there is none, but to show that the majority of firm value models does not allow for economic interpretation either. From both of these two aspects it becomes evident that the discussion in this chapter centers on FVM and we can avoid any distraction from this focus by treating RFM first.

This said, let us briefly turn to the mechanics of RFM. As was already mentioned, in these models the default event itself is solely of statistical or probabilistic nature. The determination of the default probability is exogenous to the model, as is the fixing of the loss given default. Therefore, this approach violates our criterion for economic interpretability most obviously.

The papers by Jarrow and Turnbull (1995) [48] and Duffie and Singleton (1999) [32] are standard references for models in a discrete and continuous time reduced-form framework, respectively. In their contributions, the risk of default is captured by a Poisson process whose first jump indicates default. Section 3.2 will give a more detailed treatment of this modeling idea.

A special case of RFM are rating models which have been introduced by Jarrow, Lando and Turnbull (1997) [47]. They use financial ratings that agencies like Moody's or Standard and Poor's assign to borrowers as a crucial input. Based on the history of rating changes between the categories, rating transition matrices are available from which the cumulative probabilities of default over time can be computed for a given entity.

Due to this feature, and owing to the wide-spread availability of ratings and their high degree of acceptance among investors, default probabilities derived from ratings have been considered a natural choice for the parameterization of RFM. However, because the mapping between ratings and default probabilities is based on observations of default events, these are *physical*, not *risk-neutral* probabilities. They cannot be used for pricing purposes without specifying the risk premium which adjusts the corresponding discount factor.

Unfortunately, neither is there a unique way of calibrating transition matrices to account for the market price of default risk nor is a sufficient amount of market data available from which the required risk premia could be extracted. To achieve this, there had to be traded instruments whose payoffs depend on rating changes. Such contracts have been introduced to the market only recently, but their number is far from being sufficient to calibrate the 10×10 matrix that describes the movements between the letter ratings.

Note that even a matrix of this size does not take into account the plus or minus grades of the letter ratings. It also ignores the well-known effect of rating momentum, as documented by Altman and Kao (1992) [3] and Lando and Skødeberg (2002) [50]. This refers to the phenomenon that within one rating class it is more likely for those members to be downgraded that reached this level through a downgrade than for those that were upgraded into this class and vice versa. The dimensionality of a rating model incorporating these aspects would be even higher. The reason for this is that the migration of the firms through the rating categories would become path dependent.

Obviously, the exogeneity of the default probability in RFM allows for virtually an infinite number of specifications. We will encounter more choices in the section on hybrid models (2.3.3), which follows the discussion of FVM.

2.3.2 Firm Value Models

We introduce the ideal economy of the Merton model. The framework is a frictionless market where trading takes place continuously. The riskless interest rate is constant[1] and equal to r. In this setting, a firm operates, which has the simplest of all capital structures that allow for the occurrence of default. It is financed by a single homogeneous class of debt, namely a zero-coupon bond with face value F and maturity in T, and the residual claim to the assets (equity). Assuming that the contractual agreement between the equityholders and the creditors is such that default can only occur on the maturity date T, it will occur if and only if the value of the firm's assets V_T is below F, the face value of the debt. This means that the firm is worth less than is owed to the bondholders.

A default caused by lack of liquidity is ruled out by the perfect market setting: Should the borrower's capital indeed be tied up in longer-term investments whose market value exceeds the amount owed, then it would be possible for the borrower to obtain credit from a third party. He would thereby be in the position to serve his obligations should he be forced by the original lender to do so. Naturally, the scenario that the original lender would not be willing to extend the loan could never exist in a perfect market. Hence, this is a purely hypothetical point.

---

[1] We maintain this assumption from Merton's original work for ease of exposition and notation. Relaxing this assumption does not affect the economic intuition of the Merton model.

However, the fact that the existence of a perfect market is assumed will become important later on.

Assume further (and questionably) that the dynamics of the firm's value are observable and given (under the *riskneutral*[2] martingale measure) by the stochastic differential equation (SDE)

$$dV_t = rV_t dt + \sigma V_t dZ_t,$$

where Z denotes a standard Brownian motion, σ is the volatility, and r is the (constant) risk-free interest rate. The standard Black and Scholes pricing formula for a European put can be used to obtain the price of the risky debt. Since the model's assumptions ensure the enforcement of absolute priority, the bondholders receive

$$\min(V_T, F) = F - \max(F - V_T, 0)$$

at maturity. This expression corresponds to the difference of the payoffs of a riskless zerobond and a put option on the firm's value so that the time t price of a risky bond maturing in T is

$$\bar{B}_t(T) = e^{-r(T-t)} F - P_t^{BS}. \qquad (2.1)$$

To see why the Merton model lends itself so nicely to economic interpretation, consider the two components of default risk which are reflected in the value of the put. These are the probability that the put ends in the money and the payoff from the put in this case. In the context of default risk, these quantities correspond to the probability of default and the loss given default. In the Merton model, both are determined endogenously by intuitive mechanisms. Once default has occurred in T due to an unfavorable development of the firm's asset value, it is possible to simply read off the loss given default.

However trivial this insight may seem, it turns out to be the foundation of economic interpretability. Section 2.4 will deal with extensions of the Merton model that violate the assumptions which ensure this

---

[2] The risk-neutral martingale measure is the measure obtained by choosing the riskless asset as the numeraire. The term has, of course, nothing to do with the risk attitude in the economy and is used only for convenience.

characteristic. For the sake of completeness of this introduction to default risk models, a short subsection will deal with hybrid models before we return to this question.

2.3.3 Hybrid Approaches

In some cases, it is difficult to subsume a model under the classes of RFM or FVM. Therefore, we introduce a third category of models which we refer to as hybrid. The presence of elements of reduced-form modeling typically prevents the economic interpretation of the models in this class. We will illustrate this point with two examples.

Madan and Unal (1998) [52] propose an RFM in which the probability of default depends on firm-specific information. They develop a discrete setting in which this information may take upward or downward moves at the nodes in their tree-model. The information may therefore be interpreted as the firm's value process. Although this model is often mentioned as a representative of RFM, it clearly has elements of FVM.

Zhou's paper (1997) [66] moves towards the middle ground between RFM and FVM from the opposite side of the spectrum. He adds a jump component to the process of the firm's value, so that "surprise" defaults become possible. This extension is modeled very much like the default event in RFM, namely by a Poisson process.

While it may be only a matter of taste whether one classifies these approaches as "pure" representatives of the respective model class that they try to develop further, they undoubtedly cannot be interpreted under the strict criteria we apply. In Madan and Unal (1998), the default event is not completely tied to the economic cause of bankruptcy, for instance. The default probability is a function of the firm value, but default itself is not. The exogeneity of the default event remains fundamentally unaltered despite its dependence on firm-specific information.

The setup in Zhou (1997), on the other hand, can be interpreted in a very similar way. For a given specification of the jump amplitude and the jump frequency, the process of the firm's value influences the probability of default. The difference to the paper by Madan and Unal is that this likelihood is not varied directly. Instead, the continuous part of the firm value dynamics is responsible for the indirect effect of changing the distance from the firm's value to the default boundary. Thereby, the proportion of jumps that are sufficiently large to cause default is changed without altering the probability or the magnitude of a jump itself. As such, the jump event has no economic meaning (if it had some, RFM would all of a sudden appear in a completely different light) so that it can hardly be claimed that economic intuition is present in the model as a whole.[3]

In summary, the design of hybrid models conflicts with our definition of economic interpretability. The reduced-form components are responsible for this fact. It follows that in the framework introduced firm value models are the only type of models that potentially are economically intuitive. We therefore return to this modeling approach and take a closer look at more recent representatives on this class.

2.4 Interpretability of Firm Value Models

An extensive literature has been developed to overcome the weaknesses of the original Merton model. Some of the obvious shortcomings of the approach taken in the Merton model have been successfully addressed in subsequent research. Geske (1977) [43] extended the model to the case when the firm issues a coupon bond. Default occurs, if the firm is unable to serve a coupon payment in full, so that in addition to the maturity date each coupon date represents a potential default time. In the Geske model equity is a compound option on the firm's assets.

Shimko et al. (1993) [60] and Wang (1999) [64] incorporated stochastic risk-free interest rates into the framework. The extension to debt of differing seniority is straightforward. The case of embedded options such as callability (discussed in Acharya and Carpenter (1999) [1]) is

---

[3] Note that we are not arguing that an increment of a Brownian motion, dW, can be interpreted economically while the jump of a Poisson process, dN, cannot. Both are mathematical objects that may serve as a means to convey economic intuition. The point is that the jump component dN that Zhou adds to the firm value process has a correspondence in the reduced-form setup that permits to interpret his model as a representative of this class.

complicated but does not conflict with the economic idea behind this strand of literature.

Much of the work on FVM has been motivated by the fact that it is impossible for the Merton model to reproduce the spread of risky bonds at very short maturities. Another desirable feature is to allow for the possibility of default prior to maturity.

There is no particular technical difficulty in introducing these characteristics, but doing so interferes with the praised property of economic interpretability. The literature on FVM is far too large than to allow a separate treatment of each of the proposed models. To expound on our proposition, we consider the quite general specification of Saá-Requejo and Santa-Clara (1999) [57]. It nests those of Longstaff and Schwartz (1995) [51] and Schöbel (1999) [58] as special cases. We are not aware of any model that alters the mechanics and to which our argument cannot be applied, unless the changes make it a hybrid model in our definition.

Preserving the perfect market assumption of the Merton model, Saá-Requejo and Santa-Clara follow the common idea to define the default event as the first passage time τ of the firm's value V through a so-called default boundary K. The idea of first passage models dates back to Black and Cox (1976) [11].

Allowing for a dividend payout ratio δ_v, the stochastic process for V is given (again under the risk-neutral martingale measure) by the SDE

$$\frac{dV(t)}{V(t)} = (r(t) - \delta_v)dt + \sigma_v dZ_v(t)$$

where dZ_v is the increment of a standard Brownian motion, σ_v is a positive constant and $r(t)$ denotes the risk-free short rate. This stochastic interest rate is driven by dZ_r and correlated with the firm's value via $dZ_v(t)dZ_r(t) = \rho_{rv}dt$. Several parameterizations of this process are considered in the paper. They are not relevant for the issue we are interested in.

As was pointed out before, in a perfect market, default can only occur when the assets of a debtor are insufficient to cover his liabilities. Thus, the boundary K *must* represent the value of some liabilities. The risk-neutral dynamics of K are modeled as a joint diffusion driven by the two Brownian motions dZ_r and dZ_v

$$\frac{dK(t)}{K(t)} = (r(t) - \delta_k)dt + \sigma_{kr}dZ_r(t) + \sigma_{kv}dZ_v(t).$$

The payout rate to bondholders is denoted by δ_k, and both, σ_{kr} and σ_{kv}, are positive constants. Saá-Requejo and Santa-Clara formulate their analysis in terms of the "solvency ratio" $X := \log \frac{V}{K}$. It is straightforward to obtain the risk-neutral dynamics of X from Itô's lemma. The first passage time τ is then defined by

$$\tau = \inf\{u \geq t, X(u) = 0\}.$$

Defining W ("writedown") as the loss given default, the following formula for the price of a (defaultable) bond $\bar{B}_t(T)$ issued by the firm results is obtained:

$$\bar{B}_t(T) = B_t(T) - W\, \mathbf{E}_t^Q \left[\mathbf{I}_{[\tau < T]} e^{-\int_t^\tau r(u)du} \right]. \tag{2.2}$$

In this expression, $\mathbf{I}_{[A]}$ denotes the indicator variable for event A and $B_t(T)$ is the value of the equivalent, but risk-free bond. Expectation is taken under the risk-neutral measure. Equation (2.2) produces closed form solutions for the prices of risky bonds if the chosen term structure model produces closed form prices for the risk-free bond. Moreover, it allows coupon bonds to be valued as portfolios of zerobonds because the condition for default is exogenously specified by the default boundary and not subject to strategic optimization by the equity holders as in Geske's extension of the Merton model. This reduces the computational effort considerably. Saá-Requejo and Santa-Clara point out that in their model could be extended to allow for randomness of W, as long as it is uncorrelated with r and X. In this case, W would have to be replaced by $\mathbf{E}_t^Q[W]$ in (2.2) but this possible extension is not pursued further. As we will argue in section 3.4.1 – though in a different context – this is a sensible decision. Whether W is random is not relevant for our argumentation in this chapter, either.

Apart from the elegance of its comprehensive and concise model, the paper by Saá-Requejo and Santa-Clara is also very insightful in a

2.4 Interpretability of Firm Value Models

number of ways. Most importantly, it incorporates consistently the fact that default is triggered by the insufficiency of the assets' value to cover the liabilities. The authors describe very carefully what is modeled and what is not. Nevertheless, it is instructive to examine the economic interpretability of first passage models in general and specifically that of Saá-Requejo and Santa-Clara.

Special attention must be given to the role of the writedown W. It is immediately evident that the mere existence of an exogenously specified writedown conflicts with economic interpretability according to our criterion. The loss given default is not determined within the economic scope of the model. In particular, the performance of the firm does not influence the recovery rate so that, in a way, we have already made our point. However, it may seem as if it were straightforward to endogenize the writedown. For example, introducing correlation between W and V may appear as a natural way to address this issue. But things are more complicated than that, so let us have a closer look at the problem.

First of all, in the assumed perfect market setting, such a writedown cannot exist. When the firm's value reaches the default barrier, the proceeds of liquidating the firm (which would be exactly equal to the firm's value in default) would be used to pay off the creditors (to whatever extent possible). The setting, when interpreted strictly, rules out bankruptcy costs or violations of absolute priority. Therefore, solely the value of the default boundary at the time of default should determine what lenders can recover on their commitments.

This said, however, it is easy to see why the existence of the writedown is of crucial importance for the type of first passage models. In the Merton model, it is actually the case that whatever remains of the firm's value at the default time is distributed among the creditors. There, it is due to the restrictive assumption that default can only happen at maturity that defaultable bonds trade at a discount to risk-free issues despite the absence of a writedown.

If the same default mechanism were to hold in a first passage model, creditors would receive exactly the value of K_τ were it not for the exogenous loss given default. Ignoring this feature for now throws light on the importance of the specification of the default boundary K_t. Some of the choices for this specification lead to implausible results in the perfect market scenario, keeping in mind that hitting the boundary transfers ownership of the whole firm to the creditors. For example, the constant default boundary $K_t \equiv K$ assumed in Longstaff and Schwartz (1995), favors the creditors if it amounts to the face value of total debt F or more. This is illustrated in the figure 2.1 below. When $K \equiv F$, then the firm in this example will default at time τ_1. Due to the time value of money, the earlier the default happens, the more do the lenders profit and they would actually get repaid more than they have lent out.

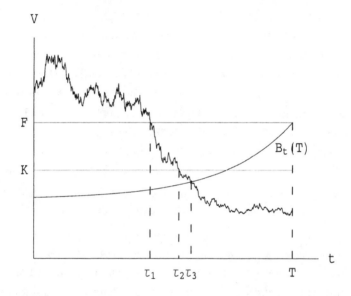

Fig. 2.1. Illustration of Possible Default Boundaries.

If – more realistically, but not guaranteed in Longstaff and Schwartz (1995) – the default boundary is below the face value, i.e. $K < F$, an early default could still be a "good deal" for the creditors. Suppose that at time τ, $B_\tau(T) < K < F$, where $B_\tau(T)$ denotes the time τ price of a risk-free bond with the same maturity (T) as the corporate bond. In this case, when the bond defaults, the lenders obtain a higher payoff on their investment, than they would have by investing in a risk-free

2.4 Interpretability of Firm Value Models

asset. In the figure, τ_2 depicts this scenario. Without loss of generality, the risk-free interest rate has been kept constant in the figure. Also, the firm value process is not terminated at the default times τ_i solely for illustrative purposes.

The general case, which encompasses the scenarios at τ_1 and τ_2, can be expressed by the condition $K_\tau > B_\tau(T)F$. An investor who receives K_τ at default of the firm and invests this sum in a default-free bond maturing in T would then have a higher terminal payoff than F. If the value of the default boundary were equal to that of a default-free, but otherwise equivalent bond, i.e., $K_\tau = B_\tau(T)F$, then the investment in a defaultable bond would be exactly equivalent to that in a risk-free one. This case corresponds to τ_3 in the graph.

Another problem when the default boundary is constant has been pointed out by Briys and Varenne (1997) [14]. At maturity, the debtors owe the complete amount F, so any threshold other than F does not make sense at this point in time. Setting K equal to F has been proven inadequate. However, if $K < F$, then the firm may not be in distress according to the default rule but all the same unable to repay its debt at maturity.

The problems attached to finding a specification of the default boundary that makes bonds actually "risky" show that the exogeneity of the loss given default is indispensable in first passage models. The root of this problem lies of course in combining the existence of a default boundary with the assumption that the firm value is observable. Suppose that it were possible to formulate a default boundary in such a way that reflected the fair value (however determined) of the total debt of the firm at all times. Rational agents would immediately liquidate the firm as soon as the firm's value hit this boundary. Given the models' assumption, they would be able to follow this strategy. Thus, they would never incur a loss.

Duffie and Lando (2001) [31] have taken up the lack of "sudden surprises" in the structural approach and borrow from T.S. Eliot to note that therefore default occurs "not with a bang but a whimper" in FVM. Clearly, this feature stands out against the typical reaction of financial markets to a default event. Duffie and Lando respond to this issue by restricting the observability of the firm value. In our scheme, the resulting model would be classified as hybrid because the paradigm of reinterpreting the discontinuities arising from this change as jumps of a reduced-form process can be applied.

2.5 Conclusion

In this chapter, we have presented a brief overview of the existing approaches to modeling default risk. We have introduced a criterion to determine whether economic interpretability can be attributed to representatives of the respective classes. In principle, only models based on a firm value approach are able to fulfill this criterion. However, the models in this category that offer a setting rich enough to reproduce even the most elementary empirical facts have to make assumptions that obliterate the economic intuition. The fully transparent and popular economic reasoning of the Merton model seems to have obstructed the view on this loss of interpretability. We conclude that from an academic perspective that values economic interpretability per se, RFM are not necessarily at a disadvantage to structural models.

In the empirical work to follow we focus on RFM for two reasons. In comparison to structural models, less attention has been devoted to this newer model class. Different FVM have been compared to one another by Bakshi (2001) [6], Eom (2002) [33], Ericsson and Reneby (2002a), (2002b) [34], [35], and Wei and Guo [65], to name only some of the most recent articles. A difficulty of implementations of structural bond pricing models is the estimation of the value and volatility of the firm's assets. In contrast to the theoretical assumptions, neither is directly observable. In estimating RFM we are able to circumvent this problem since bond data alone are sufficient.

3

Intensity-Based Modeling of Default

3.1 Introduction

Chapter 2 has introduced the most prominent attempts to modeling and predicting defaults. Since the chapter's focus has been on the clarification of the economic content of the models, reduced-form models have not been dealt with extensively. As the empirical analysis to follow will be conducted in the context of reduced-form modeling, it is the purpose of this part to make up for this shortcoming. Specifically, we will establish the theoretical foundation of intensity-based modeling and introduce theory and notation of the three specifications of this model class whose empirical performance will be scrutinized in chapter 4. Given that the ultimate goal we pursue is to test the models empirically, we emphasize the intuition behind the models. The derivations will neither be performed in full mathematical rigor nor will they encompass all conceivable variants. The reader interested in the meticulous technicalities is referred to the cited literature. The reduced-form modeling set up of the study of credit default swap premia presented in chapter 5 is very much straightforward and necessitates no separate theoretical treatment.

It has become clear that a credit spread is determined by two components: the probability of the default event and the severity of the loss given default. For a given defaultable security it is a priori all but clear which element has the greater economic influence. In terms of their respective impact on research in the field, however, the first constituent has had the greater impact by far. In accounting for this fact it becomes evident once more how structural models have shaped our understanding of default risk. Rooting the model in the stochastic differential equation for the firm value directs the focus almost automatically to the probability of reaching the default boundary. Because of the simplistic assumption we will make on the loss given default, we find it appropriate to proceed as follows. First, we define the default event and discuss its probabilistic aspects. We then devote our attention to the recovery assumption which is necessary to derive the price of a defaultable bond in general form. Thus having prepared the ground for the empirical study, we will complete the theoretical part by introducing the specifications of the models to be estimated.

3.2 Default Arrival and the Default Event

As in structural models, the default probability occupies the central role in the intensity-based environment we employ, but there is no link to the economic factors driving default. Instead, default is an event whose occurrence is determined exogenously. To gain some intuition, it is instructive to begin the discussion in a discrete setup. Beginning at time 0, it is trivial that a given issuer can either survive until t, the next point in time considered, or default. Note that we work under the physical probability measure in both this and the following section. Statistically, this single binomial outcome can be modeled using a Bernoulli distribution where

$$\Pr(default) = p(t)$$
$$\Pr(survival) = 1 - p(t)$$

and $0 \leq p(t) \leq 1$. As in the text book case of the coin toss, the binomial distribution can be derived by dividing the interval $]0,t]$ in n subintervals of equal length. Under the assumption that the outcomes are independent and the probability $p(t/n)$ is constant in each period, one obtains

$$\Pr(Y_{n,t} = k) = \binom{n}{k} p(t/n)^k \left[1 - p(t/n)\right]^{n-k}$$

where $Y_{n,t}$ denotes the number of subintervals in which "default" occurs. When the number of trials becomes large ($n \to \infty$) during the time interval $]0,t]$, and $p(t/n) \to 0$ such that $np(t/n) \to \lambda t$, where $\lambda \equiv const$, one obtains the limiting case of the binomial distribution, the Poisson distribution, with weights given by

$$\Pr(X_t = k) = e^{-\lambda t} \frac{(\lambda t)^k}{k!}. \qquad (3.1)$$

Here, X_t denotes the number of jumps ("defaults") in $]0,t]$. As such, this probability is not of much use for our purpose, since a firm can obviously only default once. A further default of an entity that came into existence through a restructuring of a bankrupt firm with the same name would be – from an economic point of view – a completely different event. What we are really interested in is the distribution of the timing of the "first" (and, should it occur, only) default. Let τ denote the arrival time of default. By definition, τ is a stopping time and thus a random variable. If τ falls in an arbitrary interval $]t_1, t_2]$, we know that the firm has survived until t_1, i.e. $X_{t_1} = 0$ but not until t_2, i.e. $X_{t_2} \neq 0$. Now we can use equation (3.1) to arrive at

$$\Pr(\tau \in]t_1, t_2]) = \Pr(X_{t_1} = 0) - \Pr(X_{t_2} = 0)$$
$$= e^{-\lambda t_1} - e^{-\lambda t_2} = \int_{t_1}^{t_2} \lambda e^{-\lambda t} dt,$$

which shows that the default arrival time is exponentially distributed with parameter λ. The parameter λ is usually referred to as the "default intensity" or the "hazard rate". Note the resemblance of the survival probabilities $e^{-\lambda t_i}$ with the price of a zero-coupon bond when the interest rate is a known constant. This similarity will carry through the following sections and will come in handy in the derivation of defaultable bond prices.

Summing up, the default event and its timing are modeled by a stopping time τ which is associated with an intensity process λ. So far, this parameter was kept constant, so that the occurrence of default was governed by a Poisson process. This assumption will be relaxed in the next section.

3.3 The Hazard Rate

Fixing the rate of default arrival at a constant level is an implausible assumption which rules out, for example, the possibility to correlate the probability of default with the overall state of the economy. Beginning with the case of a deterministically varying hazard rate, let λ_i denote the intensity prevailing in the $i-th$ time interval $\Delta_i =]t_{i-1}, t_i]$, $i = 1, ..., n$, conditional on surviving the first $i-1$ intervals. An application of Bayes' rule then yields

$$\Pr(\tau > t_i) = \Pr(\tau > t_0) \Pr(\tau > t_1 \,|\, \tau > t_0) ... \Pr(\tau > t_i \,|\, \tau > t_{i-1})$$

$$= \prod_{i=1}^{n} e^{-\lambda_i \Delta_i} = \exp\left(-\sum_{i=1}^{n} \lambda_i \Delta_i\right).$$

The equivalent expression in continuous time for the probability of survival until time $t > t_0$ is

$$\Pr(\tau > t) = \exp\left(-\int_{t_0}^{t} \lambda_s ds\right). \qquad (3.2)$$

When we generalize the setup to allow for a stochastic hazard rate, we have to form the expectation of the right hand side of equation (3.2) conditioned on the information at the inception of the assessment period:

$$\Pr(\tau > t) = \mathbf{E}_{t_0}\left[\exp\left(-\int_{t_0}^{t} \lambda_s ds\right)\right]. \qquad (3.3)$$

The existence of both, the integral and the expectation in equations (3.2) and (3.3) can be safely assumed as it will be guaranteed by a large class of processes. Note in particular that the integral is finite, and that λ must be positively bounded from zero since it is in essence a probability.

The resulting process which determines the survival probability in equation (3.3), is a doubly stochastic Poisson process (Cox process). While not a Poisson process itself, one can obtain a Poisson process from it by conditioning on the stochastic process λ_t. Once more, we would like to point out the similarity of the expression (3.3) and the a zero-bond price, this time under the regime of stochastic interest rates.

Having demonstrated that it is straightforward to extend the modeling by a stochastic process for the intensity, we turn to the second component of the credit spread.

3.4 Loss Given Default

3.4.1 Nature of the Recovery Process

As has been shown in the context of firm value models, depending on the stochastic process, the recovery rate may or may not be determined simultaneously. When recovery is specified exogenously, as is the case in the majority of models, it is hard to find a justification for varying its expected value. Bakshi et. al (2001) [6] put forward that is more natural to associate high recovery with superior credit quality but do not expound on this statement. In fact, it seems problematic to derive a forecast of the recovery rate based on the firm's value alone. There appears to be no (obvious) theoretical reason why investors should be able to recover less when the defaulting firm was on the brink of disaster compared to the scenario in which it was struck out of the blue. While it is true that the same loss which is just sufficiently large to trigger default of a financially healthy firm might wipe out a financially stricken one completely, investors may also monitor riskier firms more closely, thereby reducing the probability of losses of this magnitude. Keep in mind that credit quality usually deteriorates for some time before default occurs. It is rare that firms with an investment grade rating default on their debt.

Empirical research on this topic is extremely scarce. Data issues are an even greater problem than for defaultable bond data in general. A trivial reason for this is that default events are fortunately relatively rare. (For instance, in the sense discussed in section 3.2, any firm can only default once.) In addition, it is not an easy task to identify the recovery rate. It usually takes a long time to liquidate a bankrupt firm and this time span must be accounted for in the valuation of the remaining assets. This is virtually impossible so that post-default bond prices are taken as a proxy of the recovery rate.

The few contributions that do exist do not display an unambiguous picture. Frye (2000) [37] finds evidence in support of the hypothesis that there is a positive correlation between default probabilities and recovery rates. Altman et al. (2002) [2] present similar results. In contrast, Altman and Kishore (1996) [4] show that the original rating of a defaulting issue is irrelevant for the recovery rate, supporting the theoretical conjecture discussed above. No evidence for other possible determinants of the recovery rate has been found in this study, either. For example, the time between origination date and default date has no effect on the magnitude of the loss.

The enormous standard deviation of recovery rates within a given industry or a given rating class can explain why it is difficult to calibrate this component of default risk to actual data. For the same reason, empirical results that are obtained by using mean recovery rates only have to be treated with caution.

Apart from this, due to the additional complexity that would arise from an alternative assumption, we assume that the relative loss in case of default is exogenously given by a constant $L \in [0,1]$ so that the portion recovered amounts to $(1-L)$. This decision is compatible with our choice of the recovery regime that we describe in the next subsection. Within the chosen regime, the effects of default probability and recovery rate are not separable so that a stochastic recovery rate would make no sense at all.

3.4.2 Recovery Regime

Having decided on the (non-)stochastic nature of the recovery rate, the question of the recovery regime needs to be addressed. By this we mean the terms in which recovery is formulated. Three different notions of fractional recovery have been proposed in the academic literature. According to these proposals, recovery is either modeled as a fraction of the face value (RFV), or as a fraction of the value of a risk-free but otherwise identical bond at the time when default occurs (RT), or as a fraction of the market value which the claim had immediately before the default event (RMV). Since it is of course always possible to express recovery as a fraction of pre-default market value *ex post*, it is worthwhile to point out that the write-off under the assumption of RMV is specified before the fact, i.e. conditional on the information available at a time when the security is still alive.

We make the assumption that this last of the afore mentioned recovery regime, i.e. RMV, holds. There are well-founded arguments for the other two regimes, too. RFV, for instance, might be the best choice from a practitioner's view, considering that it is probably the most common specification in real-world contractual agreements. Modeling recovery as a fraction of a default free version appeals to the economic intuition of separating the effects of risk-free discounting and default risk. This logic is explained via a foreign currency analogy in Jarrow and Turnbull (1995) [48].

However, the chosen recovery mechanism leads to a substantial simplification of the formula for the price of defaultable bonds. For our purposes, this advantage of RMV outweighs those of the other regimes. Fortunately, the assumption is not too critical, because the deviations from the bond prices which are obtained under alternative recovery regimes are not very large as long as prices are not too different from par. This last result and the formulation of recovery in terms of pre-default market value are due to Duffie and Singleton (1999) [32]. We follow this reference in deriving the pricing formula for defaultable bonds in the next section.

3.5 Defaultable Bond Prices

As is well-known, the time t price of a non-defaultable zero-coupon bond maturing in T, $B_t(T)$, is given by

$$B_t(T) = \mathbf{E}_t^Q \left[\exp\left(-\int_t^T r_s \, ds\right) \right] \quad (3.4)$$

where r is the risk-free short rate. The expectation is taken under a risk-neutral measure, denoted by the superscript Q. In deriving the price of

3 Intensity-Based Modeling of Default

a bond that is subject to default risk the effect of the two components of default risk, the probability of default and the loss given default have to be accounted for.

Suppose for the moment that there is no recovery in case of default i.e. $L = 1$. Then, the bondholder receives a payment if and only if the default event occurs after T and the prevailing recovery regime is irrelevant. The physical probability for this scenario is given by (3.3). For valuation purposes, the expectation must be taken under Q, and amounts to

$$\mathbf{E}_t^Q \left[\exp(-\int_t^T h_s \, ds) \right],$$

where we have introduced the notation h_t for the risk-neutral hazard rate process. If we further assume independence of the processes for r_t and h_t, then the defaultable zero-coupon bond maturing in T, $\bar{B}_t(T)$, is given by

$$\bar{B}_t(T) = \mathbf{E}_t^Q \left[\exp(-\int_t^T r_s \, ds) \right] \mathbf{E}_t^Q \left[\exp(-\int_t^T h_s \, ds) \right].$$

This is simply the price of the risk-free bond multiplied by the probability of survival. Allowing for dependence between r_t and h_t, we get

$$\bar{B}_t(T) = \mathbf{E}_t^Q \left[\exp(-\int_t^T r_s + h_s \, ds) \right]. \qquad (3.5)$$

At this point we have taken advantage of the analogue functional form of survival probabilities and zero-coupon bond prices.

When L can take on non-degenerated values, the present value of any payment received in default enters the calculations and the formulation of recovery becomes crucial for retaining the simplicity of the bond price formula. To see this intuitively, consider the discrete interval $[t, t + \Delta t]$. In this setup, $h_t \Delta t$ is the conditional probability of default between t and $t + \Delta t$. The market value of the bond at t, $\bar{B}_t(T)$ consists of the sum of the present values of the recovery payment $\varphi_{t+\Delta t}$ and the survival-value of the bond $\bar{B}_{t+\Delta t}(T)$, weighted by the respective probabilities:

3.5 Defaultable Bond Prices

$$\bar{B}_t(T) = h_t e^{-r_t \Delta t} \mathbf{E}_t^Q(\varphi_{t+\Delta t}) + (1 - h_t) e^{-r_t \Delta t} \mathbf{E}_t^Q(\bar{B}_{t+\Delta t}(T)). \quad (3.6)$$

The key observation by Duffie and Singleton is that substituting the recovery of market value assumption

$$\mathbf{E}_t^Q(\varphi_{t+\Delta t}) = (1 - L) \mathbf{E}_t^Q(\bar{B}_{t+\Delta t}(T))$$

in equation (3.6) yields

$$\bar{B}_t(T) = h_t \Delta t (1 - L) e^{-r_t \Delta t} \mathbf{E}_t^Q(\bar{B}_{t+\Delta t}(T))$$
$$+ (1 - h_t \Delta t) e^{-r_t \Delta t} \mathbf{E}_t^Q(\bar{B}_{t+\Delta t}(T))$$

which can be simplified to give

$$\mathbf{E}_t^Q(e^{-R_t \Delta t} \bar{B}_{t+\Delta t}(T)) \quad (3.7)$$

where

$$e^{-R_t \Delta t} = h_t \Delta t (1 - L) e^{-r_t \Delta t} + (1 - h_t \Delta t) e^{-r_t \Delta t}.$$

Using $e^{h_t L_t \Delta t} \simeq 1 + h_t L_t \Delta t$ for small values of $h_t L_t \Delta t$, it follows that $R_t \simeq r_t + h_t L_t$. By iterating (3.7) and applying the tower law for conditional expectations, we obtain

$$\bar{B}_t(T) \simeq \mathbf{E}_t^Q \left[\exp\left(-\sum_{j=0}^{\frac{T}{\Delta t}-1} R_{t+j} \Delta t \right) \right], \quad (3.8)$$

since $\bar{B}_T(T) = 1$.

Duffie and Singleton are able to derive the continuous time analog to (3.8) as the exact relationship

$$\bar{B}_t(T) = \mathbf{E}_t^Q \left[\exp\left(-\int_t^T R_s ds \right) \right] \quad (3.9)$$

which favorably preserves the form of equation (3.4), with the default-adjusted short-rate

$$R_t = r_t + h_t L_t \quad (3.10)$$

in place of the risk-free short-rate r. In addition, the default-adjusted short-rate perspicuously reflects the presumed components of the defaultable bond price, the risk-free rate and a spread consisting of the probability of default and its consequences. The proof of equation (3.9) is spared, since it is technically involved and hardly any economic insight is to be gained from it.

3.6 Implications for the Empirical Studies

As the pricing relation developed in the previous section is fundamental for the empirical examination to come, we point out its implications for this undertaking. The form of the spread s, $s_t = h_t L_t$ implies that the two sources of default risk cannot be separated on the basis of bond data of only one seniority class. It is possible to find an infinite number of pairs $\{h_t, L_t\}$ whose multiplicative combination yields the exact same spread. In section 3.4.1 we have put forth arguments for the choice of a constant recovery rate, which were based on economic grounds and considerations of data availability. As mentioned before, this is another reason for the constant recovery assumption, which – conditioned on the available data – can be referred to as model endogenous. As we maintain this assumption, the time index of the loss rate in the expression for the default-adjusted short-rate can be dropped and (3.10) simplifies to $R_t = r_t + h_t L$. Throughout the remainder of this study, we will no longer concern ourselves with the intensity and the loss rate separately, but with the spread as a whole. Recall our assumption that the spread is entirely due to credit risk, i.e. we ignore tax effects, liquidity issues, etc.

Given the modified form of R_t, defaultable rates can be empirically analyzed in two ways. The first way is to apply a term structure model to R_t directly. As a consequence, nothing can be learned about any interaction between risk-free rates and spreads. The advantage of this method is the reduced complexity compared to the alternative two-stage procedure. This consists of first extracting the risk-free rate, usually from government securities so that the spread can then be modeled separately. We will pursue both paths in our empirical exploration. Before we are ready to suit the action to the word, however, we need to present the specifications of the interest rates whose appropriateness we wish to investigate. This is the topic of the rest of this chapter.

3.7 Affine Term Structure Models in the Context of Default Risk

Affine term structure models (ATSM) have played an important role in modeling economic variables since the publications of Vasicek (1977) [63] and Cox, Ingersoll and Ross (1985) [20]. Originally intended to model the stochastic evolution of interest rates, the stochastic processes introduced to finance in these papers have permeated practically every part of the science's quantitative side. Applications range from the field of option pricing with stochastic volatility (e.g. Stein and Stein (1991) [61] and Heston (1993) [45]) to that of modeling liquidity in Kempf (2000) [49].

The one-dimensional processes in the papers by Vasicek and Cox, Ingersoll, and Ross (CIR) are nested in the much more general class of ATSM. The general theory of ATSM has been developed in Duffie and Kan (1996) [30]. Dai and Singleton (2000) [21] provide a complete characterization of this model class, while Duffie et al. (2002) [29] fully establish its mathematical foundation.

For a number of reasons, it is quite natural to consider ATSM as candidate models for the spread. Most importantly, as in the many other applications, the choice is motivated by the numerous desirable and oft-appraised features of these models. ATSM are capable of generating a remarkable variety of shapes of term-structures while remaining expediently tractable. We will elaborate on both of these aspects below.

To use ATSM for the hazard rate is often advisable from a practical standpoint. In a perfect market setup, the return on a defaultable bond compensates the holder for the time value of the money lent to the issuer, the risk inherent in the risk-free interest rate, and the default risk. It is quite common to base default-free term structure models on the risk-free short-rate process, for example a multi-factor version of the CIR model. If the necessary infra-structure is already implemented, it will be convenient to take advantage of it when modeling the default-risky components.

3.7.1 Model Description

We now introduce the theoretical valuation framework of ATSM in a risk-free bond pricing context. For convenience, we reproduce the time t price of a zero-coupon bond maturing in T from (3.4)

$$B_t(T) = \mathbf{E}_t^Q \left[\exp\left(-\int_t^T r_s \, ds\right) \right] \qquad (3.11)$$

where r is the risk-free short rate and expectation is taken under Q. A model of the term structure is called affine if this price can be expressed as an exponentially affine function of a vector of state variables x, $x_t' = (x_{1t}, ..., x_{nt})$, i.e.:

$$B_t(\tau) = \exp(a(\tau, \xi) + b(\tau, \xi)' x_t) \qquad (3.12)$$

with $a(.)$ and $b(.)$ being a scalar and a vector function, respectively, of $\tau = T - t$, the time to maturity and some parameter vector ξ. In the sequel, the dependence on ξ is suppressed. The corresponding zero-coupon spot yield y is then an affine function of x, since

$$y_t(\tau) = -\frac{\ln(B_t(\tau))}{\tau} = -\frac{a(\tau)}{\tau} - \frac{b(\tau)'}{\tau} x_t. \qquad (3.13)$$

In ATSM, the instantaneous short rate r_t is obtained by taking limits for $\tau \to 0$ in (3.13) as

$$r_t = \delta_0 + \delta_1' x_t, \qquad (3.14)$$

where δ_0 and δ_1 are constants. Using that $B_t(\tau) = 1$ for $\tau = 0$ (i.e. at maturity) and thus $a(0) = 0$ and $b(0) = 0$ along with the definition of the derivative, we have exemplarily

$$\lim_{\tau \to 0} -\frac{a(\tau)}{\tau} = -\left.\frac{\partial a(\tau)}{\partial \tau}\right|_{\tau=0}$$

so that

$$\delta_0 = -\left.\frac{\partial a(\tau)}{\partial \tau}\right|_{\tau=0},$$

and

$$\delta_1 = -\left(\left.\frac{\partial b_1(\tau)}{\partial \tau}\right|_{\tau=0}, \ldots, \left.\frac{\partial b_n(\tau)}{\partial \tau}\right|_{\tau=0}\right)'.$$

3.7 Affine Term Structure Models in the Context of Default Risk

Duffie and Kan (1996) [30] have shown that in order to arrive at a model of affine type, the vector of state variables x_t must follow what is often called an "affine diffusion":

$$dx_t = \mathcal{K}^Q(\Theta^Q - x_t)dt + \Sigma\sqrt{S_t}dW_t^Q. \tag{3.15}$$

W_t^Q is an n-dimensional independent standard Brownian motion under Q, Θ^Q is an n-dimensional vector, $\mathcal{K}^Q$ and Σ are $n \times n$ matrices, and S_t is a diagonal matrix with the ith diagonal element given by

$$s_{iit} = \alpha_i + \beta_i' x_t, \tag{3.16}$$

where α_i is the ith element of a vector α and β_i is the ith column of a matrix $\mathcal{B}$. A canonical form for this diffusion will be presented in the next subsection.

The process for x determines $a(\tau)$ and $b(\tau)$ in equation (3.12), which have to be solved for in order to compute bond prices. In general, $a(\tau)$ and $b(\tau)$ satisfy the pair of ordinary differential equations

$$\frac{\partial a(\tau)}{\partial \tau} = -(\mathcal{K}^Q \Theta^Q)'b + \frac{1}{2}\sum_{i=1}^{N}[\Sigma'b]_i^2 \alpha_i - \delta_0 \tag{3.17}$$

$$\frac{\partial b(\tau)}{\partial \tau} = -(\mathcal{K}^Q)'b - \frac{1}{2}\sum_{i=1}^{N}[\Sigma'b]_i^2 \beta_i + \delta_1 \tag{3.18}$$

where we have suppressed the dependencies of $a(.)$ and $b(.)$ on the right hand side. The boundary conditions are $a(0) = 0$ and $b(0) = 0$. $b(\tau)$ can be solved for from (3.18) and is then substituted in (3.17) to solve for $a(\tau)$. If necessary, one can recur to rather unproblematic methods to solve the ODEs numerically. In some cases, explicit solutions are available. In particular, this property holds for the class of specifications presented in the next subsection.

3.7.2 Completely Affine Models with Independent Factors

In our empirical investigations, the attention centers on seven specifications of two- and three-factor ATSM, in which the factors follow independent Vasicek or CIR processes. Nevertheless, we introduce this model class in full generality, i.e. allowing for dependence, because we will refer to this exposition in section 3.7.4. Following the notation from Dai and Singleton (2000) [21], the specifications are labeled $A_m(n)$. In this expression, $n = 2, 3$ denotes the number of factors, whereas $m = 0, \ldots, n$ stands for the number of factors that determine the volatility structure. For instance, $A_1(3)$ would represent a three-factor model, with the first factor being driven by a CIR-type process and the other two by Vasicek processes.

Dai and Singleton provide canonical representations of admissible specifications of (3.15). Admissibility means that the conditional variances s_{iit} are strictly positive, for all i. Assume, without loss of generality, that the vector of state variables x_t is partitioned such that $x'_t = (x^{B'}, x^{D'})$, where x^B is $m \times 1$ and x^D is $(n-m) \times 1$. Then, the canonical representation defined by Dai and Singleton takes the following form (under the physical probability measure):

$$\mathcal{K} = \begin{bmatrix} \mathcal{K}^{BB}_{m \times m} & 0_{m \times (n-m)} \\ \mathcal{K}^{DB}_{(n-m) \times m} & \mathcal{K}^{DD}_{(n-m) \times (n-m)} \end{bmatrix},$$

for $m > 0$. For the case $m = 0$, a different definition applies, namely that $\mathcal{K}$ is either upper or lower triangular. Further, we have

$$\Theta = \begin{pmatrix} \Theta^B_{m \times 1} \\ 0_{(n-m) \times 1} \end{pmatrix},$$

$$\Sigma = I,$$

$$\alpha = \begin{pmatrix} 0_{m \times 1} \\ 1_{(n-m) \times 1} \end{pmatrix},$$

and

$$\mathcal{B} = \begin{bmatrix} I_{m \times m} & B^{BD}_{m \times (n-m)} \\ 0_{(n-m) \times m} & 0_{(n-m) \times (n-m)} \end{bmatrix}.$$

3.7 Affine Term Structure Models in the Context of Default Risk

Certain parametric restrictions ensure the stationarity of all processes and the positivity of those that are of the CIR type. We will use this canonical form in full generality in section 3.7.4. For now, we restrict this canonical form by requiring $\mathcal{K}$ to be diagonal and setting $B^{BD}_{m\times(n-m)} = 0$ in order to achieve independence of the components of x_t. We also set $\delta_0 = 0$ and $\delta_1 = 1$ as it has been done in much of the literature on term structure models. Freeing up these parameters allows estimation of the last $(n-m)$ elements of Θ. We can also specify Σ as a diagonal matrix with entries σ_i which provides for a more intuitive form of the processes as shown in (3.21) and (3.22) below.

The transition to the risk-neutral dynamics of the state variables is achieved by

$$dx_t = \mathcal{K}(\Theta - x_t)dt - \Sigma\sqrt{S_t}\Lambda_t dt + \Sigma\sqrt{S_t}dW_t^Q. \quad (3.19)$$

Λ_t is the market price of risk vector whose i-th element represents the price of risk associated with $W_{i,t}$. The choice for Λ implied by models of the Vasicek and CIR type is to let

$$\Lambda_t = \sqrt{S_t}\lambda_1, \quad (3.20)$$

where λ_1 is a constant n-vector[1].

Using (3.15) and (3.16) along with the restricted canonical specification, the evolution of the state variables is described by

$$dx_{i,t} = \kappa_i^Q(\theta_i^Q - x_{i,t})dt + \sigma_i\sqrt{x_{i,t}}dW_{i,t}^Q \quad (3.21)$$

for $i = 1, \ldots, m$ and

$$dx_{i,t} = \kappa_i^Q(\theta_i^Q - x_{i,t})dt + \sigma_i dW_{i,t}^Q \quad (3.22)$$

for $i = m+1, \ldots, n$. The change of measure is incorporated in the risk-neutral parameters $\kappa_i^Q = \kappa_i + \lambda_i$ and $\theta_i^Q = \kappa_i\theta_i(\kappa_i + \lambda_i)^{-1}$. Due to this choice closed-form solutions for the bond prices can be obtained. These are spelled out in section 4.2.1.2, when they enter the estimation procedure.

---
[1] The index is dispensable but kept to ensure consistency of the notation when this formula is extended by equation (3.26).

Although previous studies have demonstrated that the assumption of independence is at odds with the empirically observed behavior of the factors, we find that there are still good reasons to explore these specifications. Firstly, substantial simplification in estimating these models is due to this assumption. Not only does it reduce the number of parameters to be estimated, but it permits the use of relatively simple estimation methods. Both of these aspects can be appreciated when the specifications in this section are put in perspective to those of section 3.7.4 and when we estimate them in chapter 4. Secondly, there is a large body of literature dealing with risk-free rates that the results we obtain for defaultable rates can be compared to.

The term *completely* affine is due to Duffee (2002) [26] and motivated by the fact that in these models bond prices, as well as both, the physical and the risk-neutral dynamics, and $\Lambda_t' \Lambda_t$ are of affine type. The last property, the affine form of $\Lambda_t' \Lambda_t$, will not hold under the assumptions made in section 3.7.4. Bond prices, however, are not affected by $\Lambda_t' \Lambda_t$, which is the variance of the state price deflator, so that the specifications remain *essentially* affine.

In addition to the greater flexibility of the market price of risk within essentially ATSM, our parameterization allows for all admissible correlation structures between the state variables. Before we turn to the most general specification within the scope of this study, let us consider a simpler way to account for correlation.

3.7.3 Incorporating Correlation between Risk-Free and Risky Rates

Longstaff and Schwartz (1995) [51] and Duffee (1998) [24] have documented the inverse relationship between spreads of defaultable bonds over government securities and the level of the risk-free rate. A possible explanation for this phenomenon is that when interest rates are low, the economy tends to be in a recessionary state. Since the number of defaults is higher during this phase of the economic cycle, investors require higher compensation for default risk.

3.7 Affine Term Structure Models in the Context of Default Risk

Duffee (1999) [25] is an attempt to incorporate this stylized empirical fact. The general framework of his model is adapted from Duffie and Singleton (1999) [32]. In his setup, the default-free process is driven by two independent state variables which both follow a CIR specification. In response to the observation of correlation between spreads and the risk-free rate, the process of the intensity for firm j is tied to the default-free interest rate factors. Besides this, there is a firm-specific process to capture the stochastic fluctuation of the idiosyncratic default risk of a given firm. The relevant notation from Duffee's paper is reproduced here along with the slightly modified specification on which the corresponding empirical section is based.

The instantaneous interest rate r_t is given by

$$r_t = \alpha_r + s_{1,t} + s_{2,t}, \tag{3.23}$$

where, under Q, the processes for $s_{i,t}$, $i = 1, 2$ can be written as

$$ds_{i,t} = \kappa_i^Q(\theta_i^Q - s_{i,t})dt + \sigma_i\sqrt{s_{i,t}}\,dW_{i,t}^Q \tag{3.24}$$

and α_r is allowed to take on negative values. In fact, it is found to be negative. Due to data limitations, Duffee is unable to pin down the value of α_r exactly and sets its value to $\alpha_r = -1$ as lower values improved the fit only negligibly. The probability for negative interest rates induced by the negativity of α_r is the cost for the flexibility in fitting both steeply sloped term-structures and low, flat term-structures without imposing an unrealistic volatility behavior. Neither type of term-structure is prevalent during the sample period investigated in chapter 4, which does not overlap with the period examined in Duffee (1999). Thus, no substantial effect is to be expected by dropping the constant term, or, equivalently setting $\alpha_r = 0$ and we proceed accordingly.

For the intensity process $h_{j,t}$ of firm j, Duffee presents two observationally equivalent formulations,

$$h_{j,t} = \alpha_j + h_{j,t}^* + \beta_{1,j}(s_{1,t} - \bar{s_{1,t}}) + \beta_{2,j}(s_{2,t} - \bar{s_{2,t}})$$

and

$$h_{j,t} = \alpha_j + h_{j,t}^* + \beta_{1,j}s_{1,t} + \beta_{2,j}s_{2,t}$$

of which we choose the latter. The individual fluctuation of the spread of firm j are captured, again under Q, by

$$dh^*_{j,t} = \kappa^Q_j(\theta^Q_j - h^*_{i,t})dt + \sigma_i\sqrt{h^*_{i,t}}\,dW^Q_{j,t}. \tag{3.25}$$

The only difference between the two formulations for $h_{j,t}$ is that the constants α_j will absorb the $-\beta_{2,j}\bar{s}_{2,t}$ terms. As the increments of the Brownian motions $W^Q_{1,t}, W^Q_{2,t}$ and $W^Q_{j,t}$ are assumed to be independent, any correlation between the spreads and the interest rate is introduced by means of the firm-dependent parameters $\beta_{i,j}$. The empirical fact that even the healthiest-looking firm has a positive spread motivates the inclusion of the constant terms α_j. Notwithstanding this – from an empiricist's view – sound rationale, and although estimation actually produces positive values in Duffee's paper, the modeling seems somewhat pretentious because the intensity can still become negative. This will occur when at least one of the β coefficients is negative and its product with the respective state variable s_i is sufficiently large. This is a consequence of the general fact that the admissible correlation structure between square-root processes is restricted if the variable determined by their sum is to remain positive.

At the same time, this shows that the theoretical inconsistency is of concern only under certain parameter constellations, which is an acceptable price to pay for preserving the form of the bond price formula. Defining

$$\theta^*_i = \theta_i(1 + \beta_{i,j})$$

and

$$\sigma^*_i = \sigma_i\sqrt{1 + \beta_{i,j}}$$

and $s^*_{i,t}$ as in (3.24) with θ^*_i and σ^*_i in place of θ_i and σ_i, respectively, the price of a zero-coupon bond will be given by

$$\bar{B}_{j,t}(T) = \mathbf{E}^Q_t\left[\exp\left(-\int_t^T R^*_{j,s}ds\right)\right]$$

where

$$R^*_{j,t} = (\alpha_j + s^*_{1,t} + s^*_{2,t} + h^*_{i,t}).$$

3.7.4 Maximal Models: Essentially Affine Specifications

The most general class of models that will be estimated are the so-called essentially affine models. This class nests the completely affine class, not only for the case of independent factors as described in section 3.7.2 but also for the canonical representation provided by Dai and Singleton (2000) [21]. The distinguishing feature is a more flexible formulation of the market price of risk than that characterizing completely affine models. The model presented in this section is due to Duffee (2002) [26]. Related work in the field of extending completely affine models has been conducted by Duarte (2003) [23] and Chacko (1997) [18]. The extension proposed by Duarte leads to a semi-affine model in which the dynamics under the physical are no longer affine in x_t. Chacko's model is based on a general equilibrium approach which is substantially more complicated than the work by Duffee which we will now outline.

The definition of the market price of risk of essentially affine models necessitates the introduction of the following diagonal matrix S_t^- via its elements

$$s_{t(ii)}^- = \begin{cases} (\alpha_i + \beta_i' x_t)^{-1/2}, & \text{if } \inf(\alpha_i + \beta_i' x_t) > 0 \\ 0, & \text{otherwise} \end{cases}$$

The market price of risk in the completely affine formulation (cf. equation (3.20)) is extended to

$$\Lambda_t = \sqrt{S_t}\lambda_1 + S_t^- \lambda_2 x_t, \qquad (3.26)$$

where λ_2 is an $n \times n$ matrix. This specification increases the flexibility of completely affine models in two ways. First, the elements of the vector Λ_t can switch their sign, which is not possible in (3.20). Second, the compensation for risk in essentially ATSM is no longer proportional to the variance of the risk, thus allowing the market prices of risk to fluctuate independently of the state variables' volatility. The former feature is immediately evident from a comparison of (3.20) and (3.26). The latter can be seen best by substituting (3.26) into (3.19) and comparing the resulting physical dynamics (3.27) to those of completely

ATSM (3.28), which are obtained by substituting (3.20) into (3.19). For essentially ATSM we arrive at

$$dx_t = \mathcal{K}(\Theta - x_t)dt - \Sigma(S_t\lambda_1 + I^-\lambda_2 x_t)dt + \Sigma\sqrt{S_t}dW_t^Q, \quad (3.27)$$

where I^- is a $n \times n$ diagonal matrix with $I_{ii}^- = 1$ if $s_{t(ii)}^- \neq 0$ and $I_{ii}^- = 0$ otherwise. In the case of a completely ATSM we have

$$dx_t = \mathcal{K}(\Theta - x_t)dt - \Sigma S_t\lambda_1 dt + \Sigma\sqrt{S_t}dW_t^Q. \quad (3.28)$$

Note that in contrast to (3.28) and apart from the trivial case that $\lambda_2 = 0$, the compensation for risk in (3.27) is not proportional to the variance matrix, unless $I^- = 0$.

In general, the extent to which additional flexibility is gained in comparison to completely ATSM depends on the rank of I^- which in turn depends on the type of the processes of the components x_t. The condition $\inf(\alpha_i + \beta_i' x_t) > 0$ is only fulfilled for those components whose instantaneous volatility is not influenced by the level of the state variable, i.e. where $\alpha_i = 1$ and $\beta_{ii} = 0$ in the canonical representation. This reveals a trade-off between modeling the instantaneous volatilities of the elements in x_t and generalizing completely affine models.

The increase of flexibility is maximal for $m = 0$ because in this case, the rank of S_t^-, and hence that of I^-, is full. In general, S_t^- and I^- are of rank $n-m$, so that the first m rows of λ_2 can be set to zero without loss of generality. Thus, allowing for time-varying instantaneous volatilities by setting $m > 0$, counteracts the extension through (3.26). In fact, for $m = n$, the essentially affine formulation is identical to its completely affine counterpart.

Duffee (2002) finds that essentially ATSM have a better forecasting performance than completely ATSM. In particular, essentially ATSM can reproduce the empirically observed fact that the slope of the term structure explains a significant amount of the variation of excess returns to Treasury bonds.

The canonical form of completely affine models presented in section 3.7.2 is applicable in the context of essentially affine models as well. We use this framework for the specification and estimation of essentially affine three-factor-models in chapter 4. Therefore, the essentially affine models specified here differ from their completely affine counterparts in two ways. Besides a more flexible specification of the market price of risk, we also allow for correlation between the factors.

3.8 Summary and Outlook

In this chapter, we have given a summary of the idea and methodology of intensity-based modeling of default. We have also introduced the specific models to be estimated in the following chapter. We conclude this chapter with an overview of these models which recapitulates their main features. Furthermore, to give the reader some guidance through the following empirical part, we establish the connections to the corresponding sections of the next chapter.

Two components determine the payoff from a defaultable bond. We have simplified our setup as much as possible with respect to the loss given default by treating it as a constant. Furthermore, the decision to formulate recovery in terms of the pre-default market value of a defaultable issue simplifies the formula for the price of a default-risky bond: It is given in the same way as that of a risk-free bond but with an interest rate adjusted for default risk. This implies that using bond data, the impact of the loss given default and the default probability cannot be separated.

In this setup, two natural applications of the popular class of completely ATSM arise. We estimate a total of seven different specifications of models within this class, first using defaultable rates in section 4.2 and afterwards using spreads over the risk-free rate in section 4.3. The seven alternative specifications are obtained from the classical starter case, in which independent state variables follow either Vasicek or CIR processes, and which is described in section 3.7.2. It yields three possible combinations in a two-factor and four in a three-factor context. An overview of the total of seven different specifications is provided in section 4.2. The assumption of independence greatly facilitates the estimation of these models.

The state variables remain independent in Duffee's approach to correlate risk-free rates with spreads, as shown in section 3.7.3. To model correlation between risk-free rates and spreads is motivated by the corresponding empirical observation during the last two decades. In this model, a completely ATSM is estimated for the risk-free rate. Correlation is then introduced via the effect of the state variables that drive the risk-free rates on the firm-specific spreads. These follow a completely affine diffusion process to which the current value of the state variables of the risk-free process, weighted by some variables, are added at each point in time. Hence, the model remains in the completely affine class with independent processes. Thus, the way the feature of correlation is introduced in Duffee's model allows estimation with much simpler methods than if the state variables would be correlated themselves. In fact, it requires only slight modifications of the estimation procedure used for the calibration of the applications of completely ATSM to defaultable rates and spreads discussed before. The empirical results from Duffee's approach are presented in section 4.4.

Eventually, section 4.5 contains empirical work on the most general model in this context, which was introduced in section 3.7.4. This specification, a so-called essentially ATSM, does not only allow for correlation between the state variables, but uses a more flexible functional form of the market price of risk. The complexity of the implementation restricts us to a clinical study using defaultable rates of selected issuers.

4

The Empirical Performance of Reduced-Form Models of Default Risk

4.1 Preliminaries

In the summary of the previous chapter, we have already given a roadmap of the empirical implementations in sections 4.2 - 4.5. The estimation methods are described in these sections, too. Before we turn to the estimation of the models of chapter 3, we present the data in this section. We also discuss general data related issues and describe the estimation of risk-free and defaultable term structures using this data.

4.1.1 Data Description

The data used in this study consists of OTC Dealer quotes from REUTERS. From March 1998 until October 2001, daily quotes were collected for corporate bonds denominated in EUR and USD as well as for German and US government bonds. All estimations use midpoints calculated from these quotes. To keep the bond pricing function in the term structure estimation as simple as possible, all bonds that were non-straight in a broad sense were removed from the sample. Thus, the remaining bonds are all senior unsecured bonds that are non-callable and non-puttable, have no warrants attached and have neither variation in promised coupon payments nor sinking fund provisions. Furthermore, only bonds with a time to maturity of greater than 180 days were used, since bonds with less time until repayment of principal are highly illiquid. This becomes evident from the fact that quotes for these

44   4 Empirical Performance of Reduced-Form Models of Default Risk

bonds are very infrequent and display considerable staleness. Because the probability of default is very low over such a short horizon, removing these bonds should not have noticeable effects on the estimated default risks.

For a firm to enter the examination, it needs to have quotes for a minimum of four different issues for at least one year before July 31, 2001. This requirement is due to the number of parameters in the Nelson-Siegel procedure which is employed for the estimation of the term structure. The resulting sample for bonds denominated in EUR consists of 23 firms in total, 12 (11) of which are industrial (financial) firms. The sample of bonds with a USD denomination includes 17 firms, 5 being industrial and 12 financial firms. The intersection of the firms is made up by the 5 industrial firms in the USD sample and 10 financial firms.

Tables 4.1 and 4.2 provide detailed information on the two samples. The column labeled "number of days" provides the number of observations between the starting date and July 31, 2001, i.e. the number of days on which there where enough bond quotes available. The average number of bonds used per day is reported in the last column. It is equal to the average number of bid-ask quotes per day because if there is data for a bond on a given day, it consists of exactly one bid-ask quote. The column labeled "maturities" is explained in the next section.

While the smaller number of firms in the USD sample may come as a surprise against the backdrop of a stronger bank-oriented financing in the EUR zone, it is easily explained by the (relatively and absolutely) larger number of bonds with embedded options or other features that resulted in classifying a bond as "non-straight". The respective sample sizes are therefore not at odds with the fact that US firms rely more on capital markets as a source of funds than do European firms.

4.1.2 Defaultable Term Structure Estimation

On each day, we estimate the firm-specific defaultable term structures using the method proposed by Nelson and Siegel (1987) [54]. The spot rate curve is described by the function

4.1 Preliminaries

Table 4.1. EUR Sample

Firm	Starting date	Number of days	Maturities (in years)	Avg. bonds per day
Abbey Ntl.	Aug 27, 1998	724	3/4/6/8	7.7
ABN AMRO	Aug 31, 1999	432	7/8/9/10	5.7
Commerzbank	June 6, 1998	761	2/4/6/8	5.4
Deutsche Bank	Feb 26, 1998	838	2/4/6/8	8.6
DaimlerChrysler	Jan 6, 2000	376	2/4/6/8	5.9
Dresdner Bank	Jan 5, 1999	638	3/4/6/8	6.2
Endesa	Aug 3, 2000	188	2/4/6/8	4.6
Fiat	Feb 7, 2000	364	2/4/6/8	5.7
Ford	Aug 4, 2000	242	2/4/6/7	6.1
General Electric	Jun 8, 2000	375	3/6/8/10	20.1
General Motors	Jan 24, 2000	340	2/4/6/8	6.1
Goldman Sachs	May 18, 1999	515	3/5/7/9	4.2
HypoVereinsbank	Feb 26, 1998	842	2/4/6/8	10.8
ING	Feb 17, 1999	599	4/6/8/10	7.7
KPN	Oct 2, 2000	200	2/4/6/8	4.6
Lehman Bros.	Sep 13, 1999	445	2/4/6/8	4.9
McDonalds	Sep 6, 1999	410	2/4/6/8	4.9
Merrill Lynch	Aug 1, 2000	235	3/5/7/9	5.0
Parmalat	May 23, 2000	283	2/4/6/8	5.5
Philip Morris	Feb 26, 1998	799	2/4/6/8	5.1
Rabo Bank	Jul 7, 1999	507	7/8/9/10	11.5
Total	May 9, 2000	274	2/4/6/8	6.1
Toyota	Jun 3, 1998	686	3/4/6/8	5.5

$$r(\tau,\zeta) = \beta_0 + (\beta_1 + \beta_2)\frac{1 - e^{-k\tau}}{k\tau} - \beta_2 e^{-k\tau}, \qquad (4.1)$$

where β_0, β_1, β_2, and k are the parameters to be estimated which we collect in the vector ζ and τ is the time to maturity. Because it is a very parsimonious method in terms of the number of parameters, the Nelson and Siegel procedure can be applied when the number of bonds available for an individual corporate borrower is low. The importance

Table 4.2. USD Sample

Firm	Starting date	Number of days	Maturities (in years)	Avg. bonds per day
Abbey Ntl.	Aug 6, 1998	728	2/4/6/8	8.7
ABN AMRO	Aug 6, 1998	640 [a]	2/4/6/8	5.3
Bank of America	June 27, 2000	255	3/5/7/9	4.6
CitiGroup	Apr 6, 1999	524	3/5/7/9	5.7
Deutsche Bank	Dec 22, 1998	583	3/4/5/6	6.1
DaimlerChrysler	Mar 2, 1998	803	2/3/4/5	9.3
Dresdner Bank	Jun 19, 1998	703	2/4/6/8	6.3
Ford	Feb 27, 1998	843	3/4/6/8	13.4
General Electric	Feb 27, 1999	844	3/5/7/9	28.4
General Motors	Apr 24, 1998	805	2/4/6/8	8.9
Goldman Sachs	Jan 31, 2000	359	4/6/8/10	5.1
HypoVereinsbank	Feb 27, 1998	807	2/3/4/5	9.5
ING	Feb 19, 1999	508	2/4/6/8	4.2
Lehman Bros.	Aug 15, 2000	229	4/6/8/10	4.9
Merrill Lynch	Feb 27, 1998	832	2/4/6/8	11.5
Rabo Bank	Jan 26, 1999	617	2/4/6/8	12.7
Toyota	Jan 6, 1999	582	3/4/5/6	10.4

[a] No observations after Mar 30, 2001

of this consideration is apparent from the information on the average number of bonds per day in tables 4.1 and 4.2.

In a few cases, more sophisticated methods with a higher number of parameters could theoretically be used. Svensson's (1994) [62] extension of Nelson and Siegel would then be a natural choice. The data of some firms would even allow to fit splines. We discard these alternatives for reasons of consistency. In addition, we achieve a very good fit with the Nelson and Siegel method for all firms in our sample. We will quantify this statement shortly. This indicates that the term structures are sufficiently simple to be described by curves generated with Nelson and Siegel. Hence, there is no reason to use a different procedure.

4.1 Preliminaries

On a given day, the estimated parameters are found by minimizing the sum of squared differences between the observed market prices and the Nelson and Siegel model prices, i.e.

$$\zeta_t^{TS} = \arg\min_{\zeta} \sum_{i=1}^{N} \left(P_i^{mkt} - P_i^{mod}(\zeta) \right)^2, \qquad (4.2)$$

where N is the number of bonds on day t and the model price of the i-th bond with M payments PMT_j (of coupons and/or notional amount) is calculated as

$$P_i^{mod}(\zeta) = \sum_{j=1}^{M} e^{-r(\tau_j,\zeta)\tau_j} PMT_j. \qquad (4.3)$$

The deviation of the model prices from the observed ones are small. For almost all firms, the root mean squared error over the whole period is smaller than the average bid-ask spread. Typically, the bid-ask spread is about 25 basis points, whereas the root mean squared error is around 20 basis points. Therefore, we are quite successful in fitting the term structures to the bond prices, especially if we keep in mind that the deviations arise partly due to factors beyond the scope of any method for term structure estimation. For example, the individual bonds can only be observed non-synchronously. Bonds are not quoted as often as stocks so that non-synchronicity of the data is a more important issue.[1] Some bonds may be priced higher than others by the market because they are deemed superior in terms of liquidity related criteria such as market depth, resiliency, etc. Finally, we have ignored tax effects which could potentially distort the results.

Only in very few cases, for instance for the EUR and USD denominated bonds of General Electric, model prices are on average outside

---

[1] In addition, a purely technical issue is partly responsible for non-synchronicity. Since we have no access to the data providers tapes, bond data had to be recorded successively on every date. Due to the large amount of bonds, this process lasted throughout the whole day, so that it was not possible to restrict the time during which the observations were made. When there is access to a data base, it may be possible to select only data from, say, one hour of trading without adverse effects on the number of bonds in the sample.

the bid-ask spread. Nevertheless, the fit that we achieve is still satisfactory because the bid-ask bands are not grossly violated. For the EUR (USD) bonds of GE, an average spread of 25 (30) basis points compares to a root mean squared error of 35 (39) basis points. Since the a wider maturity spectrum is covered by the larger number of bonds for this firms, the resulting term structures are potentially more complex. However, the moderate increase in the root mean squared errors indicates clearly that the Nelson and Siegel method is still able to cope with this situation.

From each estimated firm-specific zero curve we choose four points ("measurements") representing the zero-bond yields of four maturities. Ideally, we would choose these to represent the entire term structure, i.e. beginning with a very short maturity (e.g. three months) and ending with a very long one (e.g. the maximal maturity of the bonds in the sample). The results of the term structure estimations force us to compromise on this choice. We exclude both ends of the term structures by never selecting rates for a maturity of less than two or more than ten years.

The necessity of this limitation stems from the well known fact that the functional form of the Nelson and Siegel curves can cause the estimated zero-bond yields to take on unrealistic values at the short and very long end of the maturity spectrum. In addition, the estimates for these yields are often extremely unstable over time. In this study, the uneven distribution of the data across the maturity spectrum causes the estimated curves to display both phenomena. Apart from the removal of all bonds with less than 180 days remaining until maturity, it is the clustering of the outstanding bonds of some firms around certain maturities - either temporarily or throughout the whole sample period - which contributes to this unevenness. By interpreting the cash flows in equation (4.3) as weights of the discount factors referring to certain dates, it becomes evident that the discount rate relating to the maturity date of the bond is of crucial importance for fitting the bond's price as it receives the highest weight. Hence, maturity clustering is accompanied by a concentration of weights on a set of discount rates associated with very similar tenors. At the same time, the price differences of the bonds have to be reconciled to a great extent through differences between these rates. The resulting curvature of the part of the term structure fitted to these rates may become so extreme that the Nelson-Siegel method is not flexible enough to generate a shape of the term structure that would not require the discount rates for other maturities to assume unrealistic values.

It is easy to preclude unrealistic values for the yields in question by restricting the parameters suitably. By taking the limits of $r(\tau,\zeta)$ for $\tau \to 0$ and $\tau \to \infty$ in equation (4.1), the short and long rate are obtained as $r(0,\zeta) = \beta_0 + \beta_1$ and $r(\infty) = \beta_0$. Therefore, imposing bounds on β_0 and $\beta_0 + \beta_1$ ensures that the resulting rates take on typically observed values. However, the long rate has virtually no influence on the pricing of the bonds in sample (very few of which have maturities of more than 10 years) and any estimation error of the short rate will at most have an effect on a single coupon payment - if the next coupon date is close to the observation date in the first place. On the other hand, restricting the parameters affects the treatment of all cashflows by altering the discount function altogether and therefore has a larger impact. Confronted with this tradeoff, we restrict the parameters to ensure positivity of the entire term structure but leave them unconstrained otherwise, lest to impair the fit too much. This decision reduces the number of days for which implausible term structures are estimated and enlarges the maturity spectrum from which observations can sensibly be selected. However, it cannot completely eliminate the need for restricting the maturities of the selected rates. Due to differences between the firms, the selected measurements are kept constant for each firm over the data horizon but are allowed to vary across firms. Which maturities were chosen for the individual firms is indicated in the fourth column of tables 4.1 and 4.2.

4.1.3 Risk-Free Term Structure Estimation

We need the risk-free term structures of EUR and USD interest rates for two reasons. First, we want to construct firm-specific term structures of bond spreads. Second, the model by Duffee introduced in section 3.7.3 requires as an input the factors that drive the risk-free term structure.

To estimate the risk-free term structures, we use the extension to Nelson and Siegel suggested by Svensson. The corresponding formula to equation (4.1) for the short rate of a given maturity τ is

$$r(\tau) = \beta_0 + \beta_1 \frac{1 - e^{-k_1\tau}}{k_1\tau}$$
$$+ \beta_2 \left(\frac{1 - e^{-k_1\tau}}{k_1\tau} - e^{-k_1\tau} \right) + \beta_3 \left(\frac{1 - e^{-k_2\tau}}{k_2\tau} - e^{-k_2\tau} \right).$$

The use of spline methods is quite common for the estimation of term structures for sovereign borrowers and would allow to exploit the relative abundance of outstanding debt issues. However, the Svensson method is the method of choice for the majority of the most important central banks. (Binder et al. (1999) [10] contains an overview of the methods used at various central banks.) This can be attributed to the fact that the improvement of the fit that can be achieved by using splines requires to (discretionarily) select knots, to control the smoothness of the estimated curves and to accept the resulting increase in run time.

Our results support the choice of the Svensson method. In the EUR sample, for instance, an average of slightly more than 64 bonds is used on every day to fit the term structure. The procedure results in a root mean squared error of only 0.16 EUR, compared to an average bid-ask spread of 0.066 EUR for government bonds.

4.1.4 Discussion of Data Quality

A number of data related features differentiates this empirical investigation from the related literature. To the best of our knowledge, there is currently no other study which tests the proposed models using firm specific term structures. Due to the paucity of data, it is common to use aggregated data, e.g. yield curves for certain rating categories, for empirical studies in this area of research. If firm specific data is used, it is usually limited to few bond observations. In the study by Duffee

(1999) [25], the median of the mean number of fitted bonds at each observation date is 2.47. The corresponding figures for our EUR and USD samples are 5.7 and 8.7, respectively.

Estimation and use of firm specific term structures are integral parts of this study. First of all, the process of estimating these curves is an effective means for spotting flaws in the data. Erroneous observations are easily identified as they cause massive increases in the sum of squared errors in equation (4.2).

With respect to the subsequent model estimation, the use of individual curves as opposed to single observations allows to extract information not only from the time series, but also from the cross section of a firm's bonds. Moreover, all data used for model estimation consists of zero yields, rather than par yields. Hence, the influence of differing coupon payments is limited to the indirect effect that they have on the term structure estimation. In those models, whose estimation employs a Kalman filter, the remaining effect will partly be accounted for by the estimated measurement errors, contributing to a natural interpretation of these parameters.

Compared to the time horizons commonly explored in the literature dealing with risk-free interest rates, the overall lengths of the time series considered here are rather short. In the context of risk-free rates it is indeed preferable to extent the observation period as much as possible because they typically follow slowly moving, near non-stationary processes.

When the same models are calibrated to defaultable rates, however, it must be taken into account that a corporate entity is much more likely to undergo far-reaching changes in a given time span than a sovereign borrower. The cases of Nokia, which has shifted its objective from the production of rubber boots to that of cell phones, and Goldzack, which attempted the metamorphosis from a producer of sewing supplies to an IPO-firm, may be extreme, yet excellent illustrations of this point.

As the results presented in the following sections will show, the values of the estimated parameters are of the same magnitude as those of risk-free rates. Hence, it may appear appropriate to use similar horizons for the estimation. In addition, our results document the difficulties to obtain statistically significant estimates for the means of the diffusions. As we explain in section 4.2.3.3, this would also speak in favor of a longer observation period.

Whether this could improve the results from an economic standpoint remains questionable. In any case, as Duffee (1998) [24] points out, US companies issued only few non-callable bonds prior to the mid-1980s. Thus, with any attempt to reach back further, one will encounter substantial difficulties with respect to bond valuation. Consequently, Duffee (1999) [25] uses data from 1985-1995.

We have already mentioned non-synchronous data as a potential source of error. Apart from the non-synchronicity due to the way the data is recorded during the course of each day, the quality of the data is impaired by some non-synchronous observations that result from re-scanning the REUTERS pages at the end of each day. This fills in any missing quotes that became available after the first attempt to record them. There is a trade-off between obtaining these quotes for bonds and the non-synchronicity which arises from doing so. Because firms with a comparably low number of observations would have to be dropped from the samples the consequences of re-scanning seem acceptable.

Lastly, it cannot be ruled out that some of the quotes were derived from a matrix. This means that in contrast to pricing a firm's issue individually, bond traders arrive at their quotes simply by reading them off a pricing grid which provides quotes for all combinations of maturities and rating categories within a certain range. The market prices or quotes from bonds that are perceived to be especially liquid serve as inputs for such pricing schemes. If indeed a quote comes from a matrix, it reflects an averaged price and thus conveys less information on an individual borrower. The nature of the data source is such that the existence of an active interest in trading (and therefore pricing) a particular bond can be assumed behind all quotes but is of course not guaranteed.

4.2 Estimation of Completely Affine Term Structure Models for Defaultable Rates

In section 3.7.2 the specifications completely ATSM that we will estimate in the following were introduced in the general form. We stick to the notation of Dai and Singleton (2000) [21] despite the fact that the empirical study at hand is not conducted in greatest possible generality as was pointed out before. Therefore, we refer to a specification as $A_m(n)$, where $n = 2, 3$ denotes the number of factors, whereas $m = 0, ..., n$ stands for the number of factors that determine the volatility structure, i.e. the factors following a CIR process. In total, we arrive at seven possible specifications, three two-factor and four three-factor models, which are listed in the following tables.

Table 4.3. Two-Factor Models

Specification Label	Type of Factor 1	2
$A_0(2)$	Vasicek	Vasicek
$A_1(2)$	CIR	Vasicek
$A_2(2)$	CIR	CIR

Table 4.4. Three-Factor Models

Specification Label	Type of Factor 1	2	3
$A_0(3)$	Vasicek	Vasicek	Vasicek
$A_1(3)$	CIR	Vasicek	Vasicek
$A_2(3)$	CIR	CIR	Vasicek
$A_3(3)$	CIR	CIR	CIR

The tables show that the factors are ordered so that the first m factors are the CIR factors, which are then followed by the $n - m$ Vasicek factors. We now turn to the estimation procedure.

54   4 Empirical Performance of Reduced-Form Models of Default Risk

4.2.1 Estimation Technique

One of the salient characteristics of ATSM which must be addressed in estimation is the unobservability of the underlying state variables. Basically, there are three ways to deal with this situation. Firstly, it may be possible to use a proxy. In a one-factor model for risk-free interest rates, some observed interest rate related to short maturities, for example the yield on a T-Bill maturing in three months, could serve as a proxy for the instantaneous short rate. However, for rates subject to default risk there is no equivalent instrument that one could use.

Another approach, popularized by Pearson and Sun (1994) [55], is to use the so-called "inversion method" where one selects as many observed rates as there are state variables and assumes them to be observed without error. The state variables can then be backed out for any given parameter vector using the pricing formulae. Quite unsurprisingly, the need to assume that a number of observations are not affected by any measurement error at all is subject to criticism.

The third method, which we choose in this study, overcomes this problem through the use of a Kalman filter for parameter estimation and inference of the unobserved variables. To achieve this, the dynamic systems of the ATSM to be estimated are expressed in a state-space representation. In such a formulation, the dynamics of an observed vector z_t are captured in terms of the unobserved state vector x_t as follows.

4.2.1.1 State-Space Representation

We borrow some of the notation from Duan and Simonato (1999) [22], where additional information on the use of the Kalman filter in this context can be found. Let h denote a discrete time interval, Ψ the vector of model parameters, $m(x_t, \Psi, h) \equiv \mathbf{E}(x_{t+h}|x_t)$ and $Q(x_t, \Psi, h) \equiv \mathrm{Var}(x_{t+h}|x_t)$. Then, over the time interval h, the process of the n-dimensional state vector evolves according to the equation (state, system, or transition equation)

$$x_{t+h} = m(x_t, \Psi, h) + u_{t+h}, \qquad (4.4)$$

4.2 Estimation of Completely ATSM for Defaultable Rates

where u_t is a $n \times 1$ vector assumed to be i.i.d. $N(0, Q_t)$. The observed variables z_t are linked to the state vector via the observation equation (also: measurement equation)

$$z_t = a + Bx_t + v_t. \tag{4.5}$$

In our case, the vector of measurements z_t consists of the four zero-bond yields read off the term structure on day t. Observation noise is introduced through the 4×1 vector v_t taken to be i.i.d. $N(0, R)$. Apart from the measurement error v_t and some slight notational differences, there is of course a direct correspondence between equations (4.5) and (3.13).

4.2.1.2 State-Space Specification

In ATSM, the conditional expectation $m(x_t, h)$ is affine in x_t:

$$m(x_t, h) = c(\Psi, h) + d(\Psi, h) x_t,$$

where $c(.)$ and $d(.)$ are an $n \times 1$ vector and an $n \times n$ matrix, respectively.

Since the drift term of a state variable is identical under both, the Vasicek and the CIR process, one obtains for all state variables x_i

$$c(\Psi, h)_i = e^{-\kappa_i h}$$
$$d(\Psi, h)_{ii} = \theta_i (1 - e^{-\kappa_i h}), \text{ and}$$
$$d(\Psi, h)_{ij} = 0 \text{ for } i \neq j,$$

where κ_i is the i-th diagonal element of $\mathcal{K}$.

The state variables differ with respect to their conditional variance. In the Vasicek model, the conditional variance of a state variable does not depend on its respective level. Thus, the diagonal entry q_{ii} of Q_t in equation (4.4) for a component i of x_t that follows a Vasicek process is given by

$$q_{ii} = \frac{\sigma_i^2}{2\kappa_i} \left(1 - (e^{-2\kappa_i h})\right).$$

For those components that follow a diffusion of the CIR type, it can be calculated as

$$q_{ii} = \frac{\sigma_i^2}{\kappa_i}\left(x_i(e^{-\kappa_i h} - e^{-2\kappa_i h}) + \frac{\theta_i}{2}(1 - e^{-\kappa_i h})^2\right).$$

Due to the dependence of this conditional variance on x_i, the assumption of u_t being i.i.d. $N(0, Q_t)$ is only an approximation. As is well-known from Cox et al. [20], the correct transition density is that of a non-central χ^2-distribution.

It follows from this that only for the completely Gaussian models (i.e. the $A_m(n)$ specifications for which $m = 0$) an exact likelihood function will be obtained. In all other cases, an approximate quasi-maximum likelihood estimation will be carried out. Using daily observations, this approximation can be expected to perform very well, because the deviations that can arise over such a short time interval should be rather small.

The terms in the measurement equation (4.5) are defined by the closed-form solution of the Vasicek and CIR bond pricing models for each observation $j = 1, \ldots, 4$:

$$a(j) = -\frac{1}{\tau_j}\sum_{i=1}^{m}\ln\left(\frac{2\gamma_i^{cir} e^{((\kappa_i + \lambda_i + \gamma_i^{cir})\tau_j)/2}}{(\kappa_i + \lambda_i + \gamma_i^{cir})(e^{\gamma_i^{cir}\tau_j} - 1) + 2\gamma_i^{cir}}\right)^{\frac{2\kappa_i \theta_i}{\sigma_i^2}}$$
$$-\frac{1}{\tau_j}\sum_{i=m+1}^{n}\left(\gamma_i^{vas}(B(j,i) - \tau_j) - \frac{\sigma_i^2 B^2(j,i)}{4\kappa_i}\right)$$

$$B(j,i) = \begin{cases} \frac{1}{\tau_j}\frac{2(e^{\gamma_i^{cir}\tau_j} - 1)}{(\kappa_i + \lambda_i + \gamma_i^{cir})(e^{\gamma_i^{cir}\tau_j} - 1) + 2\gamma_i^{cir}} & 1 \leq i \leq m \\ \frac{1}{\kappa_i}(1 - e^{-\kappa_i \tau_j}) & m < i \leq n \end{cases}$$

$$\gamma_i^{cir} = \sqrt{(\kappa_i + \lambda_i)^2 + 2\sigma_i^2}$$

$$\gamma_i^{vas} = \theta_i + \frac{\sigma_i \lambda_i}{\kappa_i} - \frac{\sigma_i^2}{2\kappa_i^2}$$

4.2.1.3 Kalman Filter Mechanism

The Kalman filter recursion is initiated from the unconditional distribution of the state variables. This implies the assumption of stationarity of each firm's default intensity process. The assumption is generally accepted in the literature of risk-free rates but the extension to this context may be debatable. However, given that the processes are estimated over a limited time span and the condition of all firms in the sample is quite stable during this period, the assumption seems to be acceptable in this context as well. We are therefore not forced to "sacrifice" some observations from the beginning of the time horizon in order to extract an initial distribution.

The state variables are initialized by the long-term mean:

$$x_{i,0} = \theta_i \text{ for } i = 1, \ldots, n.$$

Given these starting values of the state variables, the algorithm makes a prediction of the value of the state variables after the next time step according to the system equation (4.4):

$$x_{t-} = c + d\, x_{t-1}.$$

Analogously, starting from the unconditional covariance matrix,

$$P(i,i) = \begin{cases} \frac{\theta_i \sigma_i^2}{2\kappa_i} & 1 \leq i \leq m \\ \frac{\sigma_i^2}{2\kappa_i} & m < i \leq n \end{cases},$$

the covariance matrix of the prediction errors is predicted:

$$P_{t-} = cP_{t-1}c' + Q_{t-1}.$$

At time t, new information is learnt by observing z_t and, rearranging the observation equation (4.5), is expressed by the vector v_t of prediction errors

$$v_t = z_t - (a + Bx_{t-}).$$

The covariance matrix F of the prediction errors is given by

$$F_t = BP_{t-}B' + R.$$

Calculation of the Kalman gain K

$$K_t = P_{t-}B'F_t^{-1}$$

allows the correction of the predicted state variables via

$$x_t = x_{t-} + K_t v_t$$

and of P via

$$P_t = P_{t-} - K_t B P_{t-}.$$

Intuitively, the optimal estimates at time t are obtained by updating the best prediction available before taking the measurement with a correction term. This term consists of the deviation between predicted and measured values and is optimally weighted according to the respective variances of prediction and measurement.

A particularly appealing feature of the Kalman filter is that the contribution to the value of the (quasi-)log-likelihood function from this time step is readily available and given by

$$l_t = \frac{1}{2}(n \ln(2\pi) + \ln(\det F_t) + v_t' F_t^{-1} v_t),$$

so that by summing up these values we obtain the value of the objective function

$$\mathcal{L}_T = \sum_{t=1}^{T} l_t. \tag{4.6}$$

Duan and Simonato (1999) [22] point out that in the presence of factors following CIR processes, the assumption of normal innovations in the transition equation (4.4) is only an approximation. Therefore, the true conditional mean of the state variables differs from the recursion's prediction. The conditional variance is also misspecified because it depends on the this prediction. A nonlinear filter would be required to obtain the exact equivalence between the Kalman filter methodology and quasi-maximum likelihood. If a linear filter is used (as is the case here), then $\mathcal{L}_T$ in equation (4.6) is a quasi-log-likelihood function only

4.2 Estimation of Completely ATSM for Defaultable Rates

in an approximate sense. However, Duan and Simonato (1999) [22] provide Monte Carlo evidence on the reliability of the standard Kalman filter recursion in the context of affine term structure models. They find that the approximate quasi-maximum likelihood estimator in the case of a CIR model and the maximum likelihood estimator of a Vasicek model behave very similarly and that the asymptotic properties of both estimators hold in the finite samples as well.

4.2.2 Implementation

It is important to select the strategy for maximization of the log-likelihood function with care. Because the objective functions of these models typically possess a large number of local maxima, we employ the following two-step procedure. First, we thoroughly explore the surface of the objective function using a simulated annealing procedure. This method performs a random search which accepts changes of the parameter vector that decrease the objective function with a certain probability. The acceptance probability of a new parameter constellation is high at the initiation of the search and is reduced as the search proceeds. Thus the set of possible parameter values is narrowed. We select a relatively high initial acceptance probability and reduce it only slowly. With our setting, the number of objective function evaluations that is reached when the procedure terminates is usually around 500,000.

We then use the parameter vector resulting from this estimation as the starting values of a sequential quadratic programming method distributed by Numerical Algorithms Group. It is also available as NPSOL from Stanford University. This optimization routine is frequently proposed for the estimation of term structure and other complex models, e.g. by Duffee and Stanton (2001) [27] and Gallant and Tauchen (2001b) [41]. All routines are coded in C++.

4.2.3 Results

4.2.3.1 Preferred Models

Based on a comparison of the values of the log-likelihood functions of the two- and three-factor specifications, the preferred specification is in all cases a three-factor model. This result is robust to the use of Schwarz's Bayesian information criterion

$$BIC = -\mathcal{L}_T(\Psi) + (l_\Psi/2T)\log(T),$$

where $\mathcal{L}_T(\Psi)$ is the maximized objective function value of the right hand side of equation (4.6) and l_Ψ is the number of parameters in Ψ. The difference between the value of BIC for two- and three-factor models is in most cases substantial. Only the BIC value for the three-factor, pure Gaussian specification $A_0(3)$ is for a few firms close to, sometimes even below that of two-factor models.

Additionally, one can perform a likelihood ratio test. Note that, for instance, the $A_1(2)$ specification can be viewed as a restricted $A_1(3)$ model. It is arrived at by restricting the four parameters of one of the Vasicek processes of the encompassing specification to zero. Analogous relations exist for the other models. Then,

$$2[\mathcal{L}_T(\Psi^u) - \mathcal{L}_T(\Psi^r)] \sim \chi^2(\nu),$$

where Ψ^u and Ψ^r are the unrestricted and restricted parameter vectors corresponding to the two specifications and ν is the number of restrictions. On the basis of these tests, the null hypothesis of a two-factor model is rejected in all cases.

This finding is in line with the well-established result that three factors account for up to 99% of the total variance of the risk-free term structure. See Rebonato (1998) [56] for a textbook discussion. As the defaultable term structure comprises information on the risk-free term structure and on default risk additionally, it would have been surprising had a more parsimonious model class been identified. The result is obtained regardless of the currency denomination of the data. Within the class of three-factor models, the most successful description

4.2 Estimation of Completely ATSM for Defaultable Rates

of the data is achieved by three independent CIR factors. The results are summarized in tables 4.5 and 4.6.

Table 4.5. Preferred Models for Financial Firms

Financials	EUR	USD
Abbey Ntl.	$A_3(3)$	$A_3(3)$
ABN AMRO	$A_2(3)$	$A_2(3)$
Bank of America		$A_3(3)$
CitiGroup		$A_3(3)$
Commerzbank	$A_3(3)$	
Deutsche Bank	$A_3(3)$	$A_3(3)$
Dresdner Bank	$A_3(3)$	$A_1(3)$
Goldman Sachs	$A_3(3)$	$A_3(3)$
HypoVereinsbank	$A_3(3)$	$A_3(3)$
ING	$A_2(3)$	$A_3(3)$
Lehman Bros.	$A_3(3)$	$A_3(3)$
Merrill Lynch	$A_3(3)$	$A_3(3)$
Rabo Bank	$A_3(3)$	$A_1(3)$

Table 4.6. Preferred Models for Industrial Firms

Non-Financials	EUR	USD
DaimlerChrysler	$A_2(3)$	$A_3(3)$
Endesa	$A_1(3)$	
Fiat	$A_3(3)$	
Ford	$A_2(3)$	$A_3(3)$
General Electric	$A_3(3)$	$A_3(3)$
General Motors	$A_2(3)$	$A_1(3)$
KPN	$A_3(3)$	
McDonalds	$A_3(3)$	
Parmalat	$A_2(3)$	
Philip Morris	$A_2(3)$	
Total	$A_2(3)$	
Toyota	$A_3(3)$	$A_1(3)$

In both currency samples, about two thirds of the firms call for the $A_3(3)$-specification. For the remaining firms in the EUR sample (with the exception of Endesa), the $A_2(3)$-specification is the best model. In contrast, in the USD sample the $A_1(3)$ model is identified for the remaining firms, with the exception of ABN. There is no instance in which three independent Vasicek processes ($A_0(3)$) are singled out as the best specification for the state variables.

In the EUR sample, the results display a striking difference between financial and non-financial firms. The data of almost all firms in the former group are best described by $A_3(3)$ whereas those of industrials mostly call for an alternative specification. This result does not hold for the USD sample where the proportion of firms for which $A_3(3)$ is preferred is roughly the same for both types of firms.

An examination of the 15 firms included in both samples reveals that the process estimated for the defaultable short rate of a given firm is likely to differ across currencies. The same specification is obtained for eight firms while in the other seven cases the result is different. Interestingly, four of the five non-financial firms in the intersection of the samples are among the latter group. This finding may reflect differing characteristics of the EUR and USD credit markets that pertain to industrial firms only. Alternatively, this could be the result of a difference in the influence of the risk-free term structure on the interest rates on defaultable debt of the different types of firms. However, these conjectures must be made with the caveat that for some firms the value of the likelihood function of the preferred specification is only slightly higher than that of the second best model. In addition, the number of firms that this observation is based on is not high enough to be representative for sectors of the economy.

These objections are strengthened by the fact that we cannot identify a structural difference between the sectors in the following sections 4.3 and 4.4. On the other hand, it may simply mean that the models considered in these sections are inadequate in the sense that they are unable to pick up potentially existing differences between the respective types of firms. Although we are ultimately unable to find evidence

4.2 Estimation of Completely ATSM for Defaultable Rates

for either hypothesis, it is instructive to return to this thought in the following sections.

4.2.3.2 In- and Out-of-Sample Fit

An important diagnostic to assess the relative efficacy of the different specifications is provided by their respective in- and out-of-sample fit. The performance of the models with respect to these measures is similar in both currency samples. The representative results presented in tables 4.7, 4.8, and 4.9 were obtained from the EUR data of Deutsche Bank. They display the in- and out-of-sample fit as measured in the following way. On each day of an evaluation period, a prediction of the zero-rates for each of the four maturities is made based on the estimated parameters and the values of the state variables. The absolute deviations of these predictions are then averaged over the in-sample period and the two out-of-sample periods, respectively. The in-sample period includes the time from the firms' specific starting date until July 31, 2001. The first out-of-sample period lasts until September 10, 2001, the second covers the time until October 30, 2001. In all of the three periods, a simple random walk specification that uses the rates on a given day as the forecast of those on the next serves as a simple "straw man".

Table 4.7. In-Sample Errors for Deutsche Bank

Maturity	2	4	6	8
$A_0(2)$	0.000493	0.000364	0.000400	0.000488
$A_1(2)$	0.000493	0.000364	0.000400	0.000488
$A_2(2)$	0.000477	0.000364	0.000399	0.000485
$A_0(3)$	0.000414	0.000411	0.000448	0.000473
$A_1(3)$	0.000401	0.000382	0.000409	0.000424
$A_2(3)$	0.000383	0.000383	0.000415	0.000429
$A_3(3)$	0.000361	0.000371	0.000404	0.000401
Random Walk	0.000342	0.000365	0.000401	0.000398

Table 4.8. Out-of-Sample Errors for Deutsche Bank (1st period)

Maturity	2	4	6	8
$A_0(2)$	0.000849	0.000317	0.000333	0.000610
$A_1(2)$	0.000850	0.000318	0.000333	0.000606
$A_2(2)$	0.000786	0.000317	0.000331	0.000636
$A_0(3)$	0.000312	0.000315	0.000331	0.000356
$A_1(3)$	0.000327	0.000315	0.000324	0.000341
$A_2(3)$	0.000299	0.000316	0.000328	0.000323
$A_3(3)$	0.000324	0.000318	0.000335	0.000327
Random Walk	0.000345	0.000316	0.000324	0.000326

Table 4.9. Out-of-Sample Errors for Deutsche Bank (2nd period)

Maturity	2	4	6	8
$A_0(2)$	0.000848	0.000345	0.000347	0.000532
$A_1(2)$	0.000851	0.000345	0.000347	0.000529
$A_2(2)$	0.000766	0.000350	0.000344	0.000546
$A_0(3)$	0.000389	0.000343	0.000350	0.000364
$A_1(3)$	0.000404	0.000344	0.000338	0.000344
$A_2(3)$	0.000369	0.000345	0.000337	0.000348
$A_3(3)$	0.000384	0.000347	0.000343	0.000329
Random Walk	0.000393	0.000345	0.000339	0.000331

In terms of mean absolute errors, the data are well described by the models. A typical magnitude of this measure is about 5-10 basis points. The errors tend to be higher for longer maturities. They are also increasing when the out-of-sample period is extended. This is not a particularly surprising result because the optimal parameters normally differ over different periods. Here, however, it is of special interest because the extended out-of-sample period includes the time after September 11, 2001. Thus we are likely to also see in the results the effect of a structural break that may have occurred at this date.

When comparing the different specifications, three observations can readily be made. First, the results are very similar both in and out of the sample and across the different specifications. Second, no specifi-

4.2 Estimation of Completely ATSM for Defaultable Rates 65

cation can consistently outperform the random walk. On the contrary, the random walk has often the best fit. Third, even when ignoring the random walk, it is not possible to identify a "best" model among the seven alternatives. There is also no stable pattern of which specification is the most appropriate for a certain maturity. It is interesting, though, that the two-factor models perform much better than one would expect based on their rejection by the likelihood ratio tests and the BIC criterion.

These observations raise the issue of potential in-sample overfitting by the more complicated specifications with many free parameters. However, as Bates (2002) [8] points out, out-of-sample tests over short horizons may not be as informative as their widespread use suggests. Due to the persistency of the underlying state variables, daily zero-rates are highly serially correlated. Therefore, the potentially better forecasting ability of more complex models can often not be detected over short horizons. The results in Duffee (2002) [26] confirm this conjecture. He finds that the forecasting performance of term structure models increases with longer horizons. Owing to the limited overall time horizon of the data for this study, it is not possible to assess the out-of-sample fit over longer time spans. In addition, the discussion in section 4.1.4 concerning the possible transformation of firms over time could be a problem for such tests. Unfortunately, unlike the option-pricing literature, there is no feasible out-of-sample hedging test that could be performed with the data in this study. Hence, we lack the means of finding out whether the sophisticated models cannot be distinguished from the trivial random walk simply because of the short forecasting period.

4.2.3.3 Parameter Estimates

The estimates of the parameters all conform with prior results in the literature on the risk-free term structure. In particular, we can observe three distinctive factors in the most successful specification, $A_3(3)$. One of these is a slowly moving factor characterized by a relatively high long term mean, low volatility and low speed of mean reversion. In case of

66   4 Empirical Performance of Reduced-Form Models of Default Risk

the EUR data of Deutsche Bank in table 4.11, this is the first factor. At the other extreme, there is a factor (the third factor in the table) with a very low long term mean, substantial volatility (in comparison to the mean) and a high speed of mean reversion. The remaining factor is in between the extremes.

Table 4.10. Estimated Parameters and Standard Errors (in Parentheses) for Two-Factor Models for Deutsche Bank.

	$A_0(2)$	$A_1(2)$	$A_2(2)$
θ_1	-0.045296	0.108497	0.019423
	(0.299756)	(0.031577)	(0.012477)
κ_1	0.052321	0.464904	0.581372
	(0.000534)	(0.135644)	(0.377034)
λ_1	0.298941	-0.036727	-0.183239
	(1.54568)	(0.135458)	(0.375721)
σ_1	0.010131	0.032143	0.075754
	(0.000326)	(0.003057)	(0.00398)
θ_2	0.081077	-0.052791	0.008984
	(0.017375)	(0.117705)	(0.03504)
κ_2	0.423626	0.052098	0.302779
	(0.003181)	(0.000535)	(1.17936)
λ_2	0.341913	0.192755	-0.310179
	(0.699903)	(0.604423)	(1.17977)
σ_2	0.010586	0.010159	0.094897
	(0.001037)	(0.00032)	(0.002638)
σ_{m1}	0.000495	0.000498	0.00048
	(2.2e-05)	(2.2e-05)	(3e-05)
σ_{m2}	4.9e-05	5e-05	7.7e-05
	(4.2e-05)	(4e-05)	(2.3e-05)
σ_{m3}	0.00012	0.000119	9.3e-05
	(1e-05)	(1e-05)	(1.3e-05)
σ_{m4}	0.000363	0.000362	0.000347
	(3.4e-05)	(3.3e-05)	(3.2e-05)
LogL	-21724.7	-21722.4	-21719.6
SBC	-21643.9	-21641.6	-21638.8

The parallels to the existing research also include the magnitude of the point estimates of the parameters and, unfortunately, the difficulties in obtaining statistical significance of the estimates. While the estimates of the volatility parameters of the factors are mostly statistically significant, this does not regularly hold for the estimates of the parameters in the drift. We arrive at plausible values for the point estimates of the long-term mean, the speed of mean reversion, and the market price of risk, but achieve statistical significance only rarely.

4.2 Estimation of Completely ATSM for Defaultable Rates

Table 4.11. Estimated Parameters and Standard Errors (in Parentheses) for Three-Factor Models for Deutsche Bank.

	$A_0(3)$	$A_1(3)$	$A_2(3)$	$A_3(3)$
θ_1	-0.002784	0.020458	0.10267	0.058644
	(0.2223)	(0.085288)	(1.87329)	(3.87768)
κ_1	0.156514	0.000118	0.000173	0.003293
	(0.000811)	(0.00067)	(0.003198)	(0.217715)
λ_1	0.590132	-0.462756	-0.287948	-0.5069
	(3.55458)	(0.00445)	(0.003904)	(0.219529)
σ_1	0.00974	0.100298	0.093863	0.265066
	(0.001022)	(0.001628)	(0.001897)	(0.000385)
θ_2	-0.002979	0.047109	0.036761	0.015402
	(0.045)	(0.346632)	(0.02559)	(1.20384)
κ_2	0.604443	0.06274	0.625986	0.007203
	(0.009919)	(0.001029)	(0.438843)	(0.56239)
λ_2	0.037239	0.312839	-0.23348	-0.135812
	(1.97923)	(2.56065)	(0.438545)	(0.562498)
σ_2	0.013776	0.008488	0.056032	0.070094
	(0.000868)	(0.00043)	(0.00394)	(0.000594)
θ_3	-0.078364	-0.025863	-0.007268	0.002849
	(45.8539)	(0.072186)	(0.28186)	(0.002399)
κ_3	8.4e-05	0.421912	0.085884	1.32058
	(0.000604)	(0.003398)	(0.000894)	(1.11364)
λ_3	0.103882	0.355934	0.083654	-1.05747
	(0.412892)	(2.74324)	(2.11688)	(1.11323)
σ_3	0.00949	0.011155	0.011404	0.13839
	(0.000382)	(0.000868)	(0.00062)	(0.002131)
σ_{m1}	0.000265	0.000286	0.000206	7e-05
	(6e-06)	(1.3e-05)	(7e-06)	(1e-05)
σ_{m2}	3.2e-05	1.7e-05	2.1e-05	3e-06
	(2.4e-05)	(3.7e-05)	(2.8e-05)	(3.2e-05)
σ_{m3}	9.3e-05	6.4e-05	6.9e-05	4e-06
	(8e-06)	(8e-06)	(7e-06)	(2e-05)
σ_{m4}	0.00023	0.000183	0.000168	3.7e-05
	(1.8e-05)	(1.6e-05)	(2e-05)	(6e-06)
LogL	-22341.0	-22557.7	-22449.7	-22967.9
SBC	-22233.3	-22450.0	-22342.0	-22860.2

The estimated measurement errors $\sigma_{m1}, \ldots, \sigma_{m4}$ (the diagonal entries in the volatility matrix R of v) are small both in absolute terms and in comparison to the volatilities of the state variables. Duan and Simonato (1999) [22] obtain qualitatively similar results in a risk-free term structure estimation. Bates (2002) [8] and (2000) [7] claims that this is a typical result for options markets as well.

The lack of statistical significance of the estimated parameters has been frequently encountered in prior research. Generally, in diffusion processes, means are much more difficult to estimate than volatilities. As is shown in Campbell et al. (1997) [15], the maximum likelihood estimator of the volatility becomes more precise when the sampling interval decreases. However, more frequent sampling of the data does not reduce the asymptotic variance of the estimator of the mean. Instead, the best estimator for the mean uses as long a time span as possible. Using daily observations, we have a relatively short sampling interval, but our time-series are not particularly long. This could to some extent explain our difficulties in achieving statistical significance.

Apart from this, the standard errors may not be especially reliable for the following reason. For a sufficiently large sample size, the distribution of a maximum likelihood estimate $\hat{\psi}$ can be approximated by

$$\hat{\psi} \sim N(\psi_0, \mathcal{I}^{-1}),$$

where ψ_0 denotes the true parameter vector and $\mathcal{I}$ is the information matrix. The two ways to estimate $\mathcal{I}$ are to either compute the second-derivative estimate,

$$\hat{\mathcal{I}}_{2D} = \left.\frac{\partial^2 \mathcal{L}(\psi)}{\partial \psi \partial \psi'}\right|_{\psi=\hat{\psi}},$$

where $\mathcal{L}$ denotes the log likelihood, or to calculate the outer-product estimate

$$\hat{\mathcal{I}}_{OP} = \sum_{t=1}^{T}[h(\hat{\psi})][h(\hat{\psi})]',$$

where $h(\hat{\psi})$ is the gradient of $\mathcal{L}$ with respect to the parameter vector ψ, evalutated at the maximum likelihood estimate $\hat{\psi}$:

$$h(\hat{\psi}) = \left.\frac{\partial \mathcal{L}(\psi)}{\partial \psi}\right|_{\psi=\hat{\psi}}.$$

We are left with the second alternative because the second derivatives can neither be derived analytically nor were we able to succeed in a numerically stable implementation.[2] This means that we cannot use the White estimator for the variance-covariance matrix for $\hat{\psi}$,

---

[2] Both available alternatives, using the NAG routine that provides second derivatives directly, and applying the routine for first derivatives twice, were found impracticable with the problem at hand.

$$\mathcal{I}^{-1} = [\mathcal{I}_{2D}\mathcal{I}_{OP}^{-1}\mathcal{I}_{2D}]^{-1},$$

when the probability density is misspecified as is the case when non-Gaussian factors are present. The daily frequency of the data may mitigate this problem because the deviation from the normal density is rather small over such a short time interval.

In addition, we are confronted with numerical problems of calculating the gradients in a number of cases. We found it necessary to deactivate the automatic computation of the finite difference interval in the respective NAG routine (e04xac) and to experiment with different choices in order to arrive at meaningful results. The problem could stem from having to invert the variance-covariance matrix of the prediction errors in the Kalman filter. Matrix inversion is in general a numerically sensitive problem, and is aggravated by the small estimates that were found for the measurement errors.

4.3 Estimation of Completely Affine Term Structure Models for Spreads

It is common practice in corporate bond markets to quote prices in terms of their spread over the risk-free rate. This is a natural way to convey the pricing information because all bond prices in a given denomination incorporate the same risk-free term structure. Hence, in doing so, the focus is directed primarily on the credit risk component of the bonds' yields. The risk-free term structure can be obtained effortlessly by all market participants from the much more liquid market for government debt. The pricing problem is then to find the right markup on the risk-free rate that correctly prices the risk of default.

Our motivation to use bond spreads for the calibration of ATSM is to find out whether a superior pricing performance can be reached by separating the bond prices' risk-free and risky components. Because firm-specific term structures of spreads have not been considered at all in prior research, we can offer first insights into the complexity and characteristics of spreads, which warrants an examination of these data.

4.3.1 Implementation

We obtain the term structures of bond spreads by subtracting the estimated risk-free term structure from each firm's term structure. Then, the same four maturities as for defaultable rates are selected. Using this data, the estimation itself is carried out in the exact same way as in the case of defaultable rates.

4.3.2 Results

4.3.2.1 Preferred Models

Armed with the same diagnostics as in section 4.2.3.1 we again find that three-factor models best describe the data. This is an interesting result since unlike before, this could not be expected a priori. The finding suggests that the term structures of spreads are equally complex as those of defaultable or risk-free interest rates. Tables 4.12 and 4.13 show the exact specifications estimated for the various firms.

Table 4.12. Preferred Models for Financial Firms

Financials	EUR	USD
Abbey Ntl.	$A_1(3)$	$A_1(3)$
ABN AMRO	$A_2(3)$	$A_1(3)$
Bank of America		$A_2(3)$
CitiGroup		$A_1(3)$
Commerzbank	$A_2(3)$	
Deutsche Bank	$A_2(3)$	$A_2(3)$
Dresdner Bank	$A_2(3)$	$A_2(3)$
Goldman Sachs	$A_2(3)$	$A_1(3)$
HypoVereinsbank	$A_2(3)$	$A_1(3)$
ING	$A_1(3)$	$A_1(3)$
Lehman Bros.	$A_1(3)$	$A_2(3)$
Merrill Lynch	$A_1(3)$	$A_1(3)$
Rabo Bank	$A_3(3)$	$A_1(3)$

4.3 Estimation of Completely ATSM for Spreads

Table 4.13. Preferred Models for Industrial Firms

Non-Financials	EUR	USD
DaimlerChrysler	$A_1(3)$	$A_1(3)$
Endesa	$A_1(3)$	
Fiat	$A_1(3)$	
Ford	$A_1(3)$	$A_2(3)$
General Electric	$A_1(3)$	$A_1(3)$
General Motors	$A_1(3)$	$A_2(3)$
KPN	$A_1(3)$	
McDonalds	$A_2(3)$	
Parmalat	$A_1(3)$	
Philip Morris	$A_2(3)$	
Total	$A_1(3)$	
Toyota	$A_1(3)$	$A_2(3)$

The salient difference to the specifications estimated for defaultable rates is the greater relevance of Gaussian factors. For the spreads of EUR-denominated debt, the $A_1(3)$ model is preferred for two out of three firms. The most appropriate specification for all of the remaining firms is $A_2(3)$, except for Rabobank ($A_3(3)$). The estimated best specifications for USD spreads are either $A_1(3)$ or $A_2(3)$, with the former accounting for slightly more than half. In many cases, based on the value of the log-likelihood function, these models outperform the $A_3(3)$ version by far.

It is difficult to find an intuitive explanation for this outcome. Quite the contrary, from casual empirical observation of credit markets, where spreads seem to become more volatile when credit quality declines, one might suspect CIR processes to be more appropriate for the state variables. A similar relation between price movements and volatility can be observed in stock markets. Thus, it is entirely possible that the preference for the $A_1(3)$ and $A_2(3)$ specifications is exclusively driven by the available data and not a sensible reflection of an economically meaningful process. The parameter estimates, to which we will turn shortly, support this view.

72   4 Empirical Performance of Reduced-Form Models of Default Risk

A potential problem with the $A_3(3)$ specification when using spread data could be the low absolute value of the spreads. None of the CIR factors can take on a negative value, because the square-root in the volatility term of the dynamic of the respective factor would not be defined. Whenever a factor is set to a negative value during the Kalman filter iteration, it is therefore corrected to zero. The probability that the correction takes place is of course much higher when spread data are used for calibration since the factors must take on smaller values than in the case of defaultable rates. Therefore, the fact that we only approximate the true density of the CIR factors can have much stronger effects.

In section 4.2.3.1, we noted that for almost all industrial companies a different model was selected in the EUR and USD sample, respectively. When looking at spreads, we can no longer observe this phenomenon in such clarity. As before, the same specification for both, EUR and USD denominated debt, is identified for about one half of all firms (financial and industrial) represented in both samples. However, in the context of spreads, this ratio is now the same for both types of firms.

4.3.2.2 In- and Out-of-Sample Fit

With respect to in- and out-of-sample pricing performance, basically the same results as in the corresponding section 4.2.3.2 hold. The average absolute errors are smaller when estimating spreads. This was to be expected from the respective magnitudes of spreads and defaultable rates. In relative terms, we find that the performance of the models is better when they are calibrated to defaultable rates. It would, however, be premature to conclude from this fact that there is overfitting when applying the models to spreads because the prediction from the random walk produces higher relative errors, too.

4.3.2.3 Parameter Estimates

Once again, we can refer to the corresponding section in the context of defaultable rates, 4.2.3.3, for the discussion of general problems with the parameter estimates, although the overall level of statistical significance is higher here. The estimated measurement errors are larger than they were for defaultable rates. They are still smaller than the estimated parameters for the state variables' volatilities but large enough to obtain statistically significant estimates.

The difficulties in estimating the long-run means of the state variables remain. Taking the results at face value, we find unrealistic values for the point estimates. The estimated means of the state variable(s) following Vasicek process(es) turn out substantially negative and are not compensated by the mean parameter(s) of the CIR process(es). Despite the lack of statistical significance, this casts doubts on the meaningfulness of the preference for the $A_1(3)$ and $A_2(3)$ specifications.

4.4 Incorporating Correlation

The results of the previous section 4.3 let it appear more promising to work with defaultable rates rather than bond spreads. Hence, we return to our data set of term-structures of risky rates, this time allowing for correlation between risk-free rates and spreads. Recall from section 3.7.3 the empirical finding of a negative correlation between these quantities. Duffee (1999) [25] has introduced a model that can capture this effect without complicating the estimation too much.

4.4.1 Implementation

The implementation of Duffee's model requires separate estimation of the factors driving the risk-free term structure and of the firm-specific parameters. As shown in section 3.7.3, we adopt in principal the model in the form originally proposed by Duffee. We estimate an $A_2(2)$ model for the risk-free short-rate. In this step, we can recur to the exact same procedure as in section 4.2. In contrast to Duffee, we set $\alpha_r = 0$ in equation (3.23). We comment on this choice further below. A second minor change is to use the second of the two observationally equivalent formulations of the idiosyncratic spread process proposed by Duffee. For the estimation of this process, some straightforward adjustments in the Kalman filter formulae must be made according to the setup described in section 3.7.3. After that, estimation and implementation can be carried out analogously to the procedure described in section 4.2.

Since our main motivation to consider Duffee's model is to gain an understanding of the relevance of correlation effects, we do not compare different specifications within this approach. The extension of the model to a higher number of factors driving the risk-free rate and/or the spread would only distract from this purpose.[3] Thus, when we present the results of the estimation in the following section, we focus on the in- and out-of-sample fit and the parameter estimates.

4.4.2 Results

We have two benchmarks to compare our results to. One is the original research conducted by Duffee. Apart from the fact that his data naturally differ from ours in terms of the firms included and the time horizon, there are some structural differences between the two data sets. The first is that the observations in his sample are of monthly frequency, unless there are missing observations in which case it is even longer. This is a substantial deviation from the daily frequency of our data. Secondly, Duffee cannot use firm-specific term structures in the estimation since the number of observations for each firm on a given data is not sufficient to derive them. Instead, the yields of the bonds are used directly. As a consequence, coupon effects may materialize in the results, as was discussed in section 4.1.4. Moreover, the interpretation of the measurement errors of the Kalman filter procedure is obscured. The dimension of the measurement equation for the firms is actually not clear in Duffee (1999). The measurements themselves must relate to time-varying maturities over the observation period. Hence, one cannot associate the estimates of the measurement errors (which are not reported in Duffee's article) with certain maturities. This makes it difficult to gain an insight from these parameters.

---

[3] We also avoid the problems of trying to fit a large number of factors to a small quantity that we encountered in section 4.3.

The comparison to Duffee's results is also impaired by our slightly different formulation. However, when we follow Duffee in setting $\alpha_r = -1$, the value of the likelihood function is substantially lower for two out of three firms. For the remaining third of the sample, the value is typically only barely higher. Even more damning is the substantially worse in- and out-of-sample fit that results from setting $\alpha_r = -1$. Therefore, we accept the loss of comparability due to our choice of $\alpha_r = 0$. As argued in section 3.7.3, the much more homogenous shape of the risk-free term structures throughout our observation period does not necessitate the constant.

The results of our estimation of completely affine models for default-risky rates in section 4.2 constitute the other benchmark. In both setups we use three factors to model default-risky term structures. In contrast, section 4.3 is not directly comparable, since there we use already three factors to describe the spreads. The total number of factors used for modeling defaultable term structures resulting from this separation into risk-free rates and spreads would be higher than three due to the additional factors of the risk-free term structure.

4.4.2.1 In- and Out-of-Sample Fit

The quality of the fit in the estimation and the control period varies widely, both by firm and maturity. The fluctuations of the fit tend to be more pronounced for the USD data. To illustrate this finding, we use the results for the USD rates of Dresdner Bank, which appear in table 4.14.

By all comparisons, the model achieves a good fit of the zero-rate for the maturity of four years. Within the estimation period, the mean absolute error is about 3.7 basis points, almost 1 basis point lower than the corresponding random walk error. The basic ATSM models of sections 4.2 and 4.3 produced only rarely such an outperformance of the random walk. In the two out-of-sample periods, the forecasting ability is very close to that of the random walk. Considering the short forecasting period, Duffee's model is quite successful. Surprisingly, over the longer period of 12 weeks, it even outperforms the random walk.

76   4 Empirical Performance of Reduced-Form Models of Default Risk

Table 4.14. In- and Out-of-Sample Errors for Dresdner Bank (USD)

Maturity (years)	2	4	6	8
In-Sample Fit				
Mean Absolute Error	0.001428	0.000368	0.000760	0.001672
Random Walk Absolute Error	0.000544	0.000447	0.000511	0.000622
Out-Of-Sample Fit over 6 Weeks				
Mean Absolute Error	0.002019	0.000597	0.001875	0.003636
Random Walk Absolute Error	0.000496	0.000562	0.001042	0.001894
Out-Of-Sample Fit over 12 Weeks				
Mean Absolute Error	0.001977	0.000595	0.001709	0.003083
Random Walk Absolute Error	0.000530	0.000616	0.001145	0.002017

For the other maturities, especially those of two and eight years, the performance is much worse than that of the random walk. For example, the out-of-sample forecasting error for the two-year rate is about four times as high as that of the random walk. Such extreme variations of the fit across the different maturities were an exception in the context of the models considered before.

Also, we have so far not observed such pronounced differences in the fit across the firms as become apparent for the model at hand. While the results for Dresdner Bank are representative for many firms regardless of the currency denomination, some firms show completely different results. For instance, in the EUR sample, the rates of General Electric of all maturities are forecasted more accurately by Duffee's model over both out-of-sample periods. The results for the EUR rates of Ford are at the other extreme. In this case, one can hardly speak of a fit at all. For the longest maturity, the forecast is off by over 130 b.p.

Duffee provides only aggregated information on the in-sample fit. For each firm, he calculates the root mean squared error of yield to maturity as the square-root of the mean of the squared differences between the actual and fitted yields to maturity on the firm's bonds. He reports the median (9.83 b.p.) and the 1st (7.39 b.p.) and 3rd (11.05 b.p.) quartile of this measure. Although these figures are not directly

4.4 Incorporating Correlation

comparable to our results due to the aggregation, the magnitude of the in-sample errors seems to be of the same order as our results. If we take the average of the in-sample errors relating to the four maturities we obtain a value of 10.57 b.p. for Dresdner Bank.

If we use this statistic to compare the in- and out-of-sample fit of our implementation of Duffee's model to that of the starter cases in the previous sections, we find that the basic ATSM are superior in terms of the fit. In the case of Dresdner Bank, we arrive at an average in-sample error of 6.15 b.p. for the preferred $A_1(3)$ specification. This specification, as is the case with many specifications preferred by the BIC, does not even have the best fit. Therefore, this is a conservative comparison. The failure of Duffee's model at the short and long maturities is responsible for this fact.

4.4.2.2 Parameter Estimates

In table 4.15, we briefly present the estimated parameters of the two-factor square-root model calibrated to the risk-free term structure. We need this information to interpret the estimates of the correlation parameters between the two factors and the idiosyncratic spreads of each firm. For the reasons stated above, our results differ from the estimates in Duffee (1999), but some similarities remain. As Duffee, we find two distinguishable factors. Due to the higher mean, the first factor can be thought of as a factor representing the level of the curve. We find higher estimates for the volatilities, but at least for the first factor this is compensated by a higher parameter of the speed of mean reversion.

Tables B.1, B.2, and B.3 in the appendix display the parameter estimates for the spread processes of the individual firms. It is hard to derive insights from the estimated values for the constants α_j and the parameters of the process of the individual fluctuations given by equation (3.25), not least because only the estimate for the volatility was consistently statistically significant. Overall, these estimates are very similar to those found by Duffee, who also reports great uncertainty in the parameter estimates.

78   4 Empirical Performance of Reduced-Form Models of Default Risk

Table 4.15. Parameter Estimates of the $A_2(2)$-Model for the Risk-Free EUR and USD Short-Rates

	EUR		USD	
i	1	2	1	2
θ_i	0.0160	0.0063	0.0282	0.0065
κ_i	0.5917	0.3521	0.4222	0.1914
λ_i	-0.2559	-0.3930	-0.0423	-0.3232
σ_i	0.0878	0.1514	0.0695	0.1383

However, the most interesting parameters are undoubtedly the correlation parameters β_i because they distinguish the model from those considered in previous sections. The estimates obtained are statistically significant and with few exceptions negative. This means that in line with empirical evidence spreads are negatively correlated with the risk-free rate in this model. Note that since the first risk-free factor (the level factor) is much larger than the second, a negative estimate for β_1 outweighs a positive value for β_2. The estimates for both parameters tend to be higher in absolute value for the USD data. Since both risk-free factors have higher means, this indicates a higher degree of (negative) correlation for this currency.

Let us take up once more the comparison between financial and industrial firms motivated by the results in section 4.2.3.1. One might expect the degree of correlation between the risk-free rate and the spreads to differ between the two types of firms for two reasons. The first is that banks rely much more heavily on debt financing than industrial firms. Secondly, when the term structure is upward sloping as is usually the case, they can earn substantial profits from term transformation. This strategy, which is also known as "riding the yield curve", involves investing in assets with a longer maturity than that of the refinancing debt. If the distinctly different capital structure and profit drivers caused a stronger correlation for firms in the financial sector, we should see diverging estimates for the correlation parameters β_i. Actually, we do see drastically varying values for these parameters, but within both groups so that financial and non-financial firms cannot be distinguished from these estimates. However, against the backdrop of the mixed overall performance of the model, we cannot conclude that there is indeed no difference between the defaultable rates in the respective sectors.

4.5 Estimation of Essentially Affine Term Structure Models for Defaultable Rates

Having investigated a number of possible specifications in the context of intensity-based default modeling, we must record that none of the alternatives appears undisputably as the optimal choice. As in the ongoing research on the risk-free term structure, a natural direction for research is to experiment with more flexible model designs. This motivates the study of essentially ATSM, which we present in this section.

Unlike completely ATSM, models of this type are not straightforward to estimate. Due to the correlation between the state variables and the more flexible form of the market price of risk, essentially ATSM possess an intractable likelihood function. This problem is not only encountered in the term structure literature but also present in multi-factor option pricing models, for example. In response to this difficulty, it is natural to consider approaches that combine simulation with moment matching. The drawback of these techniques is their computational intensity which forces us to limit our investigation to a clinical study.

The method which has evolved as the standard for estimation in these cases is the efficient method of moments (EMM) introduced by Gallant and Tauchen (1996) [38]. Our discussion here draws mainly on the subsequent and more refined publications by Gallant and Tauchen (2001) [39], (2001a) [40], (2001b) [41], and (2002) [42].

4.5.1 Estimation Technique: Efficient Method of Moments

The basic idea of the EMM estimator is as follows. First, the information of the empirical data is exploited by estimating an auxiliary model for this data. The auxiliary model describes the conditional density of the time series of observations and is used to generate the data-dependent moment conditions. It does not have to have anything to do with the structural[4] model one is ultimately interested in. All that is to be achieved by the estimation of the auxiliary model is a good statistical description of the data. Second, a long time series is simulated from the structural model. The parameters of the structural models are found by matching the scores of the auxiliary model using the simulated path.

Formally, let $\tilde{y}_t$ denote the vector of yields observed at time t. Let the history of these observations through t be summarized in $\tilde{Y}_t = (\tilde{y}_1', \ldots, \tilde{y}_t')'$. (The tilde denotes a specific outcome of the random variable y_t.) The auxiliary model then expresses the density of y_t as a function of Y_t and a parameter vector ρ:

$$f_t(y_t|Y_{t-1}, \rho).$$

The parameter vector ρ is estimated with Quasi-Maximum Likelihood:

$$\tilde{\rho}_T = \arg\max_\rho \frac{1}{T} \sum_{t=1}^T \ln f_t(\tilde{y}_t|\tilde{Y}_{t-1}, \rho), \qquad (4.7)$$

where it is assumed that the form of $f_t(.)$ has been chosen such that this estimation is tractable. The standard choice for the auxiliary model to use with EMM is the SNP (Semi**N**on**P**arametric) model of Gallant and Tauchen (2001) [41]. This is also the model that will be used in this study. Theoretically founded in the Hermite series expansion, it is possible to achieve an arbitrarily close approximation to any smooth conditional density with this method. The density function is defined by an Hermite polynomial $\mathcal{P}$ multiplied by a normal density $\phi(.)$,

$$f_t(y_t|Y_{t-1}, \rho) = C[\mathcal{P}(z_t, Y_{t-1}^L)]^2 \phi(y_t|\mu_{Y_{t-1}^L}, \Sigma_{Y_{t-1}^L})$$

where $\mathcal{P}$ is a polynomial with degree K_z in z_t, which is a normalized version of y_t defined by

---

[4] In this context, the term "structural" is used in a different sense that is not to be confused with its meaning in chapter 2. There, it referred to firm-value modeling of default. Here, it sets apart the model to be estimated – which usually possesses an *economic* structure – from the potentially purely statistical auxiliary model.

4.5 Estimation of Essentially ATSM for Defaultable Rates

$$z_t = R^{-1}_{Y^L_{t-1}}(y_t - \mu_{Y^L_{t-1}}).$$

In this expression $\mu_{Y^L_{t-1}}$ is the mean and $R_{Y^L_{t-1}}$ is the Cholesky decomposition of the variance of y_t conditioned on $\tilde{Y}^L_{t-1} = (\tilde{y}'_{t-1}, \ldots, \tilde{y}'_{t-L})'$, the vector of the L lags of observed data. $\mathcal{C}$ is a constant of proportionality equal to $1/\int[\mathcal{P}(z_t, Y^L_{t-1})]^2 \phi(s) ds$.

A number of tuning parameters allows for great flexibility in fitting the conditional distribution. Typically, the choice of parameter settings is arrived at by first determining the number of lags in the VAR mean specification, L_u. Then, the order of the ARCH (L_r) and GARCH (L_g) specification of the variance of y_t is chosen. Non-Gaussian behavior can be introduced by increasing the dimensionality K_z of the Hermite polynomial. These coefficients are allowed to be another polynomial in Y^L_{t-1} of degree K_y. This is another way to introduce conditional heterogeneity. However, it results in a nonlinear, nonparametric specification. The lag order L_p on the polynomial of the coefficients is inoperative as long as $K_y = 0$. It is set to one only by convention. The restrictions implied by settings of the tuning parameters are summarized in table 4.16.

For example, in a Gaussian GARCH model, the transition density of y_t conditioned on previous observations is normal. The conditional mean at time t is given by a VAR model, the conditional variance by a GARCH model. The number of lags used in calculating the mean is denoted by L_u. L_r and L_g provide the number of lags in the autoregressive and moving-average part of the model for the variance, respectively. Conditional on the information from the VAR and GARCH model, the density is normal, hence the label "Gaussian". If this conditional normal density is multiplied by an Hermite polynomial ($K_z > 0$), then the density has an additional non-parametric component which makes it – in the terminology of Gallant and Tauchen – semiparametric model, in this case a semiparametric GARCH model.

The specification of an SNP-type auxiliary model is characterized by an eight-digit "code" that is made up by the six digits in table 4.16 plus two additional digits I_z and I_y for the number of cross product terms suppressed in K_z and K_y, respectively. Table 4.17 in section 4.5.3.1 provides exemplifications.

Table 4.16. Restrictions Implied by Settings of the Tuning Parameters

Parameter Setting	Characterization of y_t
$L_u = 0, L_g = 0, L_r = 0, L_p > 0, K_z = 0, K_y = 0$	iid Gaussian
$L_u > 0, L_g = 0, L_r = 0, L_p > 0, K_z = 0, K_y = 0$	Gaussian VAR
$L_u > 0, L_g = 0, L_r = 0, L_p > 0, K_z > 0, K_y = 0$	semiparametric VAR
$L_u > 0, L_g = 0, L_r > 0, L_p > 0, K_z = 0, K_y = 0$	Gaussian ARCH
$L_u > 0, L_g = 0, L_r > 0, L_p > 0, K_z > 0, K_y = 0$	semiparametric ARCH
$L_u > 0, L_g > 0, L_r > 0, L_p > 0, K_z = 0, K_y = 0$	Gaussian GARCH
$L_u > 0, L_g > 0, L_r > 0, L_p > 0, K_z > 0, K_y = 0$	semiparametric GARCH
$L_u \geq 0, L_g \geq 0, L_r \geq 0, L_p > 0, K_z > 0, K_y > 0$	nonlinear nonparametric

The criterion for all of these decisions is Schwarz's Bayesian information criterion

$$BIC = s_T(\tilde{\rho}_T) + (l_{\tilde{\rho}_T}/2T)\log(T),$$

where $s_T(\tilde{\rho}_T)$ is the negative maximized objective function value of the right hand side of equation (4.7) and $l_{\tilde{\rho}_T}$ is the length of the auxiliary model. The search strategy for finding an appropriate parameterization is discussed in more detail in Gallant and Tauchen (2001b) [41].

After specifying the auxiliary model, one can obtain the score vector associated with that model,

$$m_T(\tilde{\rho}_T) = \frac{1}{T}\sum_{t=1}^{T}\frac{\partial}{\partial \rho}\ln f_t(\tilde{y}_t|\tilde{Y}_{t-1}, \tilde{\rho}_T),$$

which provides the moment conditions in the following simulated moment matching estimation procedure. For this reason, the auxiliary model is also called the score generator.

Using a candidate parameter vector ψ of the structural model, a long time series $\hat{Y}_N(\psi) = (\hat{y}_1(\psi)', \ldots, \hat{y}_N(\psi)')'$ is simulated from the structural model. The score vector of the auxiliary model is reevaluated using the simulated data,

$$\hat{m}_T(\psi, \tilde{\rho}_T) = \frac{1}{N}\sum_{t=1}^{N}\frac{\partial}{\partial \rho}\ln f_t(\hat{y}_t(\psi)|\hat{Y}_{t-1}(\psi), \tilde{\rho}_T),$$

4.5 Estimation of Essentially ATSM for Defaultable Rates

and used in the estimator of ψ,

$$\hat{\psi}_N = \arg\min_{\psi} \hat{m}_T(\psi, \tilde{\rho}_T)' \tilde{I}_{OP,T}^{-1} \hat{m}_T(\psi, \tilde{\rho}_T),$$

where $\tilde{I}_{OP,T}^{-1}$ is the outer-product estimate of the auxiliary model's information matrix

$$\tilde{I}_{OP,T}^{-1} = \frac{1}{T} \sum_{t=1}^{T} \left[\frac{\partial}{\partial \rho} \ln f_t(\tilde{y}_t | \tilde{Y}_{t-1}, \tilde{\rho}_T) \right] \left[\frac{\partial}{\partial \rho} \ln f_t(\tilde{y}_t | \tilde{Y}_{t-1}, \tilde{\rho}_T) \right]'.$$

An estimate of the asymptotic variance-covariance matrix of the estimator is given by

$$\hat{\Sigma}_T = \frac{1}{T} \left[(\hat{M}_T)' \tilde{I}_{OP,T}^{-1} (\hat{M}_T) \right]^{-1},$$

where

$$\hat{M}_T = \frac{\partial}{\partial \psi} \hat{m}_T(\psi, \tilde{\rho}_T)$$

are the gradients of the moment conditions. Assuming that the number of moment conditions l_ρ is greater than the number of structural parameters l_ψ, the overidentifying restrictions can be tested by using that

$$T \hat{m}_T(\psi, \tilde{\rho}_T)' \tilde{I}_{OP,T}^{-1} \hat{m}_T(\psi, \tilde{\rho}_T)$$

is asymptotically distributed as a $\chi^2(l_\rho - l_\psi)$ random variable.

4.5.2 Implementation

We obtain our results with the code for the SNP model and the EMM procedure provided by Gallant and Tauchen. The programs are coded in Fortran 77 and are publicly available from an ftp server at Duke University. Naturally, all data input related procedures have to be adapted to the data. In addition, the structural model has to be coded by the user for the simulation during the estimation step. Since our essentially ATSM generates the unobservable vector of state variables whereas the score generator fits observable yields, we also have to compute the yields resulting from the state variables. The yields in the dynamic term structure models that we explore cannot be computed in closed form, so we have to solve the ODEs (3.18) and (3.18) numerically. As is common in the literature, we use the Runge-Kutta method, here as implemented by NAG in routine d02pdf, for this task. We have checked the accuracy of the ODEs via monte carlo simulations.

Data availability forces us to restrain ourselves to a clinical study. Above all, the number of observations for the estimation of the auxiliary model is not high to begin with. We identify four firms, all of which are banks, for which we have comparably long time series in both currencies. When we estimate the auxiliary model we find it necessary to reduce the number of observations in the cross-section. Instead of the four rates we have worked with before, we use only two, namely the shortest and longest maturity that were read off the term structure. Due to the huge increase of the number of parameters, we were unable to arrive at meaningful estimates for the parameters in our attempts to use a four-dimensional observation vector. Consider for instance, that when using two lags the VAR component of the auxiliary model requires 36 parameters in the four-dimensional case, compared to 10 for two dimensions. Still, as shown below, the preferred specifications of the auxiliary model require about 40 parameters in total. For most of these, however, statistically significant estimates can be obtained.

To cope with the problem of multiple local optima, the following feature of the SNP/EMM implementation is used. For the first trial, the program imputes start values. In the following trials, these are perturbed by multiplying them by $(1+w)q$, where w is a uniform[-1,1] random variable and q is a scaling factor that controls the size of the perturbation. The perturbed start values are passed to the optimizer (NPSOL), which carries out a certain (user-specified) number of iterations. We repeat this step 10 times and use the best result as the start value for further optimization. The whole procedure is repeated for various levels of the scaling factor q, ranging from 10^{-5} to 10^0.

4.5.3 Results

4.5.3.1 Auxiliary Model

As table 4.17 shows, we obtain quite similar preferred specifications for the four firms in both currencies. The first six digits of the specification-

4.5 Estimation of Essentially ATSM for Defaultable Rates

"codes" are in the same order as in table 4.16. The additional two digits are the values of I_z and I_y, the number of cross product terms suppressed. For example, the transition density of the EUR time series data of Abbey National is – using the labels in table 4.16 – a semiparametric GARCH. The first component of the density, the conditional normal has two lags in the VAR specification for the conditional mean, and one each in the autoregressive and moving average component of volatility. Departures from conditional normality are modeled by an Hermite polynomial of order four, the second component of the density which accounts for the non-parametric part.

Table 4.17. Preferred Auxiliary Model Specifications

	EUR	USD
Abbey Ntl.	21114000	31114000
Deutsche Bank	21114000	31114000
Dresdner Bank	30114000	40214000
Rabo Bank	30014000	20114000

The order of the Hermite polynomial is the same for all firms. In fitting this part of the auxiliary model to the data, we trust the experience-based advice of Gallant and Tauchen "never to consider a value of $K_z < 4$", which they give in the user's guide to their SNP implementation [41]. When the conditional density does deviate from normality, $K_z = 4$ is thus the most parsimonious choice. The specifications of the ARCH and GARCH component are quite parsimonious as well. With respect to the conditional mean, we observe that USD data tends to require more lags in the VAR part. In no case is a non-parametric specification called for, as $K_y = 0$. The second to last digit (equal to 0) indicates that no cross-product terms are suppressed in the Hermite polynomial. The last digit provides the same information on the polynomial in the coefficients but is inoperative since $K_y = 0$.

4.5.3.2 Structural Model

Unfortunately, we are rather unsuccessful in the estimation of the structural models for any of the four firms. Table B.4 in the appendix contains the estimation results for the EUR data of Deutsche Bank but this information has to be treated with much caution. We have great difficulties in achieving convergence in the nonlinear optimization. It is therefore not surprising at all that we encounter the problem of achieving statistical significance again and that all models are rejected on the basis of the χ^2-statistic. Due to these difficulties, we do not pursue the estimation of essentially ATSM any further.

Most of the problems we encounter are anything but new in applications of the EMM methodology. Gallant and Tauchen themselves call attention to many of them in their manual for EMM [40]. Among the possible reasons for the failure are problems with the dynamic stability of the score generator and/or of the underlying structural model, misspecification of the score generator, unfavorable small sample properties of the methodology, start value problems, and numerical problems. We devote the rest of this section to address these issues.

Dynamic stability refers to the long-run properties of a dynamic model. Simulations from both the score generator and the structural model must be nonexplosive. Then, the conditions of stationarity and ergodicity are fulfilled. According to Gallant and Tauchen, one must be especially attentive to this issue when using levels data like nominal interest rates, particulary when fitting diffusion models to such data, which is obviously the case in our study. False local optima are to be expected as a consequence of dynamic instability. In response to this issue, Andersen and Lund (1997) [5] use an E-GARCH specification instead of a GARCH model to ensure dynamic stability. We rely on the suggested spline transformation feature in the SNP user's guide [41] which was not yet available to Andersen and Lund. Gallant and Tauchen claim that this transformation enforces dynamic stability on the score generator. They have also obtained the result, that a dynamically stable score generator enforces stability on the underlying structural model.

4.5 Estimation of Essentially ATSM for Defaultable Rates

More importantly, our results could be affected by misspecification of the score generator. Monte Carlo evidence from Zhou (2001) [67] demonstrates that for highly persistent data, up to seven (ten) different SNP specifications are required to describe the transition densities of 89% (96%) of the simulated samples of size 500. For less persistent data, one or two specifications are sufficient. This evidence, albeit obtained using an older version of the SNP method, suggests that the approximation of the transition density to that of the true data generating process may not always be reliable. However, the whole estimation procedure rests on the assumption that the true transition density is closely approximated.

Duffee and Stanton (2001) [27] investigate the small sample properties of the EMM methodology for a two-dimensional time-series simulated from a standard CIR model. The first dimension is the simulated short rate, the second the one year yield calculated from the short rate with white noise added to prevent perfect correlation. They find that the parameter estimates using EMM are biased and highly dependent on the start values. The χ^2-statistics turned out to be overstated. Despite its theoretical inconsistency, an approximate Kalman filter produced the best results in their study. The authors conclude that EMM behaves poorly in samples typically used for term structure estimation and advocate the use of a Kalman filter.

Gallant and Tauchen are aware of the potential difficulty in getting decent start values. In the EMM user's guide [40] they suggest "intensive use" of the program's feature to generate randomly perturbed start values. This strategy involves running "many" trials using different starting points for perturbations of varying magnitude. The run-time of about three to four days on a Pentium IV 2GHz machine under Linux, which was required for evaluating one set of starting values perturbed 50 times, gives an idea of the numerical intensity of the procedure. The length of the simulation of the structural model was set at 50,000, which is typically considered sufficient, but not excessively long. Some efficiency may be gained by first evaluating the parameter constellations with shorter simulation runs and using only the most promising settings with an increased length afterwards. We observe widely differing results with different start values. Much more experimentation would have been required to rule out the possibility that we simply have the wrong start values, but this is precluded by the excessive run-time required for this task. This is also one of the major reasons why we were not able to estimate the models for more firms.

Finally, numerical problems impede the process of estimation. Gallant and Tauchen discuss the necessity of what they call "bullet proofing" the data generating process, i.e. in our case the simulator of the essentially ATSM. The need arises from the tendency of the optimizer (NPSOL) to try "outlandishly extreme values" of the parameter vector ρ of the structural model. These values can result in causing numerical exceptions (dividing by zero, taking the logs or the square roots of negative numbers) within the simulation, the evaluation of the score in the SNP procedure, or the optimizer itself. In this context, "bullet proofing" refers to ensuring that a sensible outcome of the simulation is guaranteed independent of ρ. The user is powerless when these problems occur in the SNP procedure or within NPSOL. While the executable files created under Linux are at least able to deal with the exception and continue to run, we observe that NPSOL often fails to recognize such combinations as bad choices for the maximization of the objective function. In addition, the procedures that carry out the Runge-Kutta method to solve the ODEs also struggle with extreme parameter vectors. It is sometimes not possible to evaluate the ODEs to a satisfying accuracy.

In light of the preceding discussion, it appears evident that much work remains to be done before models of the complexity of essentially ATSM can be estimated with EMM for defaultable data. It would be interesting to carry out further research to assess the relative efficacy of different approaches to estimation. For example, whether the alternative estimator advocated by Duffee and Stanton could work in this context is an open question. The prerequisite for this would be that essentially ATSM are actually able to fit the empirical features of defaultable rates. Unfortunately, we have not come close to answering these questions.

4.6 Summary

The nonuniform and sometimes contradictory results and the difficulties to obtain them in the first place are apt to raise the question of what can be learned from the material presented in this chapter. Indeed, having presented a considerable amount of empirical work, it is about time to draw some conclusions.

First, note that inhomogeneity of the results can convey in itself an important insight. For example, the notable differences with respect to the preferred specification of completely ATSM in the base case confirms the notion of default risk as a firm-specific risk. The varying performance in terms of in- and out-of-sample fit and differences across currencies and industries are supportive of this view. Although pooling of data across rating classes and/or industries may be the only feasible alternative for practical purposes, this insight underlines the importance of further investigating this risk on firm-specific data.

Regarding this task, the results suggest that there is room for improvement in the formulation of default risk models. Although very small absolute errors could be obtained under some of the specifications, a simple random walk is all too often a superior alternative. As discussed before, one should not ignore the possibility that this outcome is simply due to the restrictive tests we were able to conduct with our specific data. This is another issue that calls for further investigation. Nevertheless, a certain degree of misspecification of the existing models is to be suspected.

However, the clinical study of essentially ATSM has clearly shown that at this point, data availability is insufficient to implement the more sophisticated models of the kind currently used in research on the risk-free term structure. In this context, the specific difficulties in the estimation via EMM in addition to the problems already encountered in the estimation of the completely ATSM models posed an insurmountable hurdle.

We postpone further conclusions to the final chapter of this book when we are able to include results from the shift of focus in the next chapter. We study data from a separate, but closely connected market. Using a simpler methodology, we investigate credit default swap premia.

5

Explaining Credit Default Swap Premia

5.1 Introduction

In the previous chapters, we have already noted that extant literature on the pricing of claims whose payoff is strongly determined by default risk still displays a substantial lack of empirical work. There is a slowly growing literature on the pricing of defaultable debt, but hardly any empirical contributions have been made using credit derivatives data. Even the pricing of the most common credit derivatives, plain-vanilla credit default swaps (henceforth CDS) has up to now not been investigated empirically except for Houweling and Vorst (2001) [46]. This chapter, based on Benkert (2004) [9], contributes to filling this gap in academic research.[1] Before the backdrop of continuing growth of the market for CDS and their increasing use in credit-linked structured products, the practical importance of this question is quite obvious. In fact, the International Swap Dealers Association (ISDA) reports the outstanding volume of credit derivatives as of year-end 2002 to have reached $2 trillion.

We pursue a second strand of research in that we explore the effect of equity volatility on the market's evaluation of default risk. Recent publications by Campbell et al. (2001) [16] and Goyal/Santa-Clara (2002) [44] have brought to mind the importance of idiosyncratic volatility, which so far has received surprisingly little interest in the context of

---
[1] The chapter closely follows the forthcoming article. It has been adapted to blend in with the other chapters but the content is mainly identical.

empirical research on default risk. By including option-implied volatility in the analysis – in addition to historical volatility – this idea is taken up and developed further.

The application of any model of default risk to CDS is complicated inasmuch as "traditional" models are usually primarily concerned with the pricing of risky debt as opposed to valuing a credit derivative like a CDS. Although the connection between corporate bond spreads and CDS premia is close enough to justify the use of similar models and methodologies, one must take the difference between these quantities into account in one way or another. This can be achieved by deriving a model for that difference. Such a task may well be too ambitious – at least in terms of calibration to actual data and practical use. Or, alternatively, one can model CDS directly, in which case the implementation of the models reviewed below is likely to be infeasible and/or inadequate.

In chapters 2 and 3, the different attempts that have been made to model the pricing impacts of default risk were introduced. We have categorized them into firm value (or structural), hybrid, and reduced-form models, the latter group including rating transition models. We briefly recall the specific difficulties that implementations of such model bring about to motivate our setup.

The fundamental problem of firm value models is the unobservability of the input parameters to this equation. For example, in the simplest case, represented by the original Merton model (1974) [53], the spot firm value and its volatility are unobservable. Typically, researchers resort to proxies for these quantities. This is increasingly difficult, the less active the trading in a firm's securities. Structural models often fail to produce a realistic spread for higher-rated and short-term debt. On the other hand, lower-rated firms tend to have more complicated capital structures that set hurdles for the implementation of this model class. In a comparison of firm value models by Eom et al. (2002) [33], the simplicity of the capital structure is an important criterion for sample selection.

5.1 Introduction

The calibration of intensity based models based on CDS premia directly is prevented by the paucity of data, at least for meaningful specifications of the hazard rate process. The history of CDS is simply not long enough to provide time series of the lengths required to fit stochastic processes reliably. And despite the rapid growth of the market in recent years, for a considerable number of obligors, time series are still frequently interrupted by periods of missing observations. Moreover, even CDS data for the most frequently traded reference entities occasionally display substantial staleness.

Pure rating-transition models require the specification of a mechanism to transform the physical migration probabilities supplied by rating agencies into risk adjusted probabilities. The number and type of traded securities does not yet allow the extraction of this information in a unique way from market data. Additionally, the problem of selecting (or modeling) an adequate recovery rate must be addressed separately.

As for a purely statistical, possibly non-parametric fitting of CDS premia, the lack of data of sufficient quality is again prohibitive. Apart from leaving an economist uncomfortable, the total loss of structural insights and theoretical foundation obstructs the use of any results when trying to price instruments for more illiquid reference assets.

A possible solution to the pricing problem that avoids the model-specific issues outlined above while modeling CDS premia directly, may be found in conducting regression analyses of the kind used, for example, by Collin-Dufresne et al. [19] and, most recently, Campbell and Taksler (2002) [17]. These authors try to identify and measure the determinants of corporate bond spreads. The advantages of their suggested model are twofold. The economic structure is kept at a level which appears to be appropriate for the amount and quality of available data. In addition, the use of panel data mitigates specific shortcomings of the data, for instance gaps in the borrower-specific time-series. In this chapter, we adopt the modeling idea of Campbell and Taksler and extend their analysis in a variety of aspects.

The salient contributions to empirical investigations of default risk that result from our approach are the following. First of all, our research is one of the very few studies conducted using CDS data. Secondly, in the spirit of Campbell and Taksler, we give special emphasis to the repeatedly neglected issue of the link between equity volatility and credit spreads. The major contribution in this respect is to not only consider various measures of historical volatility, but to include option-implied volatility in our analyses. Furthermore, using CDS data, it is possible to enlarge the spectrum of rating classes beyond the scope typically considered in empirical research. The data also facilitates the investigation of a more international set of firms than could be achieved by concentrating on corporate bonds.

The remainder of the chapter is organized as follows. Section 5.2 introduces the model and establishes the theoretical foundation for its application to CDS data. We present the data in section 5.3 and define the explanatory variables. The estimation procedure and the results are reported in section 5.4, followed by robustness checks in section 5.5, and the conclusion in section 5.6.

5.2 Modeling Idea

Following Campbell and Taksler, we undertake a regression analysis that requires less economic structure than a typical representative of a full-fledged firm value model. Nevertheless, it is firmly rooted in the theory of structural modeling of default risk and incorporates available rating information as well.

It is the inclusion of equity volatility that connects the statistical model with structural modeling of default by adopting the following logic from Merton (1974) [53]. An investment in a corporate bond can be perceived as the combination of a riskless bond with the short-sale of a put on the issuing firm's value. Thus, an increase in volatility will shift probability mass toward the tails of the distribution of firm value. Economically speaking, this raises the probability of default. In terms of the replicating portfolio, this economic fact is reflected by an increase

5.2 Modeling Idea

of the price of the put, worsening the position of the creditors. In such a scenario, rising volatility is therefore accompanied by rising spreads and vice versa. Note that this argument does not depend on the market price of risk.

Having described the central point of the setup employed by Campbell and Taksler, we complete this outline by reviewing briefly the remaining variables. In order to avoid redundancies, both their relevance in the context of CDS data and – if applicable – their respective economic interpretation will be discussed in section 5.3.

All regressions conducted by Campbell and Taksler include proxies for the term structure of risk-free interest rates, liquidity, and bond-specific features. The regressors they use to examine default risk consist of financial ratios, rating information, and measures of market-wide as well as firm-level equity risk and return. Controlling for variation in the cross section of firms and in the time series using fixed effects, Campbell and Taksler examine the explanatory power of the respective proxies by including them in different combinations in their regressions.

The rationale for applying the basic scheme of the study conducted by Campbell and Taksler to CDS data is founded on the theoretical insight that under idealized conditions the premium the protection buyer has to pay in a CDS agreement equals the spread offered by a risky bond offers over a riskless, but otherwise equivalent, instrument.

To understand this, consider the contractual specification of a CDS. In a CDS agreement the so-called protection buyer is entitled to a payment from the counterparty (the protection seller) that compensates him for a loss on the principal of a reference asset due to default of the issuer. In return, the protection seller receives periodic payments from the insured party. As in a "classical" swap, this CDS premium is agreed upon such that no money changes hands when the contract is entered. From an economic standpoint, the strategies of either investing in a riskless bond or buying a defaultable bond and protecting its proceeds through the purchase of a CDS on the applicable reference asset should allow for no arbitrage gains. Thus, the protection seller should receive compensation for assuming the default risk of the reference asset's issuer and the CDS rate should be set equal to the spread of the defaultable bond over a riskfree but otherwise identical bond.

Discussing the necessary assumptions in more detail, Duffie (1998) [28] shows that the equivalence of the two strategies holds exactly only under the regime of recovery of market value for a defaultable floater that currently trades at par. These requirements are obviously not fulfilled by typical corporate bonds since firms usually issue fixed rate debt or floating rate debt with a fixed spread over Libor. Furthermore, while the assumptions market participants make about the recovery mechanism when pricing corporate debt are unknown and quite difficult to extract from bond prices, it may be argued that the concepts of recovery of treasury and recovery of face value are more intuitive and for this reason more popular, too.

However, it is shown by Duffie that the deviations that result from these violations are rather small. They are not sufficiently large to explain the empirically observed divergence of bond spreads and CDS premia completely. These differences can instead be attributed to the existence of market imperfections. In his starter case, Duffie assumes the costlessness of reverse-repo arrangements and the absence of transactions costs and tax effects. This list would have to be amended so as to rule out price impacts resulting from liquidity issues and regulatory capital arbitrage. In this case, the argument of Duffie continues to hold in principle. The magnitude of the divergence is relatively small compared to the level of both the CDS premia as well as the bond spreads. Hence, CDS premia represent primarily a price of default risk and are in this respect similar to bond spreads. Consequently, CDS premia and bond spreads should be driven by the same factors. This provides the motivation to proceed analogously with the regression analysis.

Beyond the validity of the procedure, a number of distinct advantages arise when using CDS data instead of bond spreads. Firstly, one need not correct for maturity in the regressions. The available data suggest a natural restriction to CDS with a maturity of five years, which is the most frequently traded maturity by far. Moreover, since all quotes are for newly initiated contracts, all observations in the sample have the same maturity.

Secondly, in contrast to bonds, there are no coupon effects to be concerned about. CDS, like all swaps, are designed so that no cash flow occurs at initiation of the contract. Put differently, dealing with CDS quotes is comparable to having a sample of corporate bonds that trade at par on each and every day. Issue size is another bond specific variable that is not applicable in the context of CDS quotes which relate to a borrower as an entity. Differences in liquidity among several bonds of one debtor are therefore not an issue.

Thirdly, the analysis of CDS premia offers the chance to extend empirical research in the area of default risk to non-investment grade issuers. Bonds rated below BBB are usually excluded from the data, as is also the case in the work by Campbell and Taksler. Their data represent the trading activity of insurance companies for whom high-yield debt is "particularly unattractive" due to reserve requirements faced by that industry. Insurance companies may even be prohibited from buying such bonds in the first place. Consequently, Campbell and Taksler discard all non-investment grade debt transactions in their data as unrepresentative of the market as a whole. The main reason for the exclusion of junk bonds from other empirical work in the field lies in the fact that a majority of these instruments are callable, puttable, or convertible, or equipped with other features such as sinking-fund provisions and step-up coupons. To avoid the pricing difficulties induced by these characteristics and to preserve the homogeneity of the sample, these bonds are typically deleted from the samples, causing lower rating classes to be underrepresented if not completely ignored. Note that a sample bias will be introduced among higher rated firms, too, if the extent to which these firms use non-straight bonds to raise capital differs greatly.

Lastly, as discussed earlier, using CDS data in the first place naturally avoids the problem of having to calibrate information gained for bond spreads to CDS premia, if one is interested in modelling the latter. Of course, this comes at the cost that the factors that are responsible for differences between these two quantities might distort the coefficients of the independent variables. It can, however, not be ruled out that this would not be the case if the regression was run using bond spreads. Confronted with the two alternatives, the second seems to be the lesser evil.

5.3 Data

The data used in this chapter consist of quotes on US dollar denominated CDS on senior unsecured debt. All contracts have a maturity of five years. The market for credit derivatives is still entirely an OTC market for which no centrally assembled data base exists. This study's data were collected and made available to us by WestLB. It consists primarily of indicative bid and ask quotes in this market. The depth and transparency of the CDS market is far from being comparable to that of mature markets for equity or interest rate derivatives. It has been argued that for this reason quotes may be unreliable and transaction data should be used instead. However, a transaction is made either at the bid or the ask quote, which instead introduces errors such as the bid ask bounce when the spread between the buy and the sell side is large. This is often the case in a young market like that for CDS. In this study, the problem is alleviated to some extent since for many entities there often exists more than one quote on a given day. Whenever this is the case, a "market spread" is constructed by combining the most favorable buy and offer quotes. The mean and median of the bid and ask spreads in the sample are 16.44 and 9 basis points, respectively. All equity and accounting data were downloaded from Bloomberg. The interest rate information is publicly available from the US Federal Reserve Board.

The sample includes CDS spreads for 120 international borrowers from various industrial sectors and spans the period between January 1, 1999 and May 31, 2002. Altogether, 26478 quotes (midpoints generated from the spreads) are used in the estimations. The number of observations per firm varies significantly between a minimum of 19 to a maximum of 516. Within this range, it is fairly evenly distributed.

5.3 Data

Descriptive statistics on the variables are presented in table 5.1. Since the accounting variables change quarterly at most, analogous information is not reported for them. Two variables are included in all of the regressions to capture the general conditions of the economy and to control for liquidity. Corresponding to the maturity of the CDS in the sample, we use the 5-year treasury yield as a proxy for the level of the risk-free term structure. The exact choice of the level proxy was found not to be important. We do not use a proxy for the slope, because the shape of the term structure did not display much variation during the sample period. By constructing a slope proxy as the difference between a long and a short rate, the risk of multi-collinearity would have been introduced. Duffee (1998) [24] has documented the inverse relationship between spreads of defaultable bonds over government securities and the level of the risk-free rate. An explanation for this empirical fact that is often brought forward as a theoretical underpinning associates low interest rates with a recessionary state of the economy. Corporate defaults occur more often during economic downturns than during boom phases and the occurrence of a recession may cause a decline in credit quality that leads to more defaults in the future. According to this line of reasoning the compensation for default risk would rise and a negative sign of the coefficient would be expected.

Table 5.1. Descriptive Statistics of Independent Variables.

Variable	Mean	Std Dev	Minimum	Maximum
Implied Volatility	42.03	16.48	10.02	180.74
Historical Volatility	40.62	18.07	11.78	170.63
Gov5y	5.11	0.73	3.57	6.69
Liquidity	0.0003	0.0035	-0.01680	0.08134

In addition to the interest rate, a variable is needed to control for variation in the demand for liquidity during the sample period. Following the terrorist assaults of September 11, 2001, a "flight to quality" was observed in the financial markets, and we try to capture this effect by including the difference between the 3-month USD interest rate swap and Treasury yields. A widening of this difference indicates decreasing confidence of the market participants in the stability of the banking system. For this variable, a positive coefficient is to be expected because in times of stress, when the price of a standard swap above the government bill rises, investors will require a higher price to accept the default risk of corporate bonds.

The effect of the respective rating classes is measured relative to a rating of AA. There are no CDS in the sample that refer to higher rated entities, firms rated worse than B, or unrated firms. Dummy variables are included for all rating classes below AA that are represented in the sample. Ratings are reported using Standard and Poor's notation and reflect the average of the respective Standard and Poor's and Moody's ratings, or whichever rating was available if the company was not rated by both organizations. As for the predictions for the coefficients, it is to be expected that they are positive and inversely related to the quality of the respective borrower. This follows directly from the properties of the rating systems.

To proxy for past profitability, leverage, and interest coverage, the financial ratios of EBIT to net sales, long-term debt to total assets, and EBIT to interest expense are included. The economic interpretation of these components is straightforward. A firm which is able to sell its products successfully is more likely to repay its debt. Hence, the coefficient for the first ratio should take a negative sign as investors would require less of a spread for money lent to successful marketers. When the burden of debt is rather high, the firm may default no matter how profitable it is in terms of turnover. This motivates the inclusion of a measure for leverage in the regression further establishing the linkage to the fundamental logic of firm value models. Recall that the amount of money owed by a company represents the strike price of the put in the duplicating portfolio for the corporate debt. Other things equal, the value of the put is an increasing function of the strike. Since the strike price must be put into perspective it enters the regression scaled by a measure of the capacity to bear debt. The coefficient for the resulting leverage proxy, long-term debt to total assets, should be positive. The third ratio, EBIT to interest expense, combines the effects of profitability and leverage to the extent that the numerator of the profitability ratio is scaled by a quantity closely related to the numerator of the leverage ratio. In addition, it is affected by the interest rate payable on the debt. Since a low interest coverage may portend the firm's failure to meet its obligations even before maturity of the debt, a negative sign for the coefficient is to be expected.

The accounting information is (hopefully) tied to the economic performance of the firms in the sample. This raises the question of whether its informational content is already reflected by the rating assigned to the respective borrowers. Rating agencies do use accounting data but it is not known exactly which kind and in which way. They also consider a significant number of other factors in their evaluation of credit quality and judge creditworthiness based on an analysis "through the business-cycle" rather than on a snapshot. Thus, in contrast to accounting data, ratings are not entirely backward looking. It follows from this that neither ratings nor accounting information are redundant, a priori. In order to learn about the relation between firm characteristics and CDS premia, both types of variables should enter the regression.

With the data at hand, it is impossible to separate the effect of equity volatility into a systematic and idiosyncratic component. As opposed to the sample examined by Campbell and Taksler, this would require data for not just one but many stock indices and open up the issue relative to which benchmark excess returns for each individual firm should be measured. Although this loss of information is regrettable, it does not affect the investigation of the impact of equity volatility on CDS premia, since the relevant quantity is evidently total volatility. It enters the regressions as the annualized standard deviation of daily equity returns over 180 calendar days prior to the date of the respective quote.

Further, we include the Black-Scholes volatility implied in at-the-money stock options in our regressions. In doing so, we do not intend to rehash the discussion about the usefulness of implied volatility as a forecast of future realized volatility or as an estimator of equity volatility which has repeatedly been led in a logically inconsistent way. It is a fallacy to use implied volatility as an estimator for equity volatility if one assumes the Black-Scholes model to hold, because in this case volatility must be treated as a constant and hence be derived from past observations. If one does not believe in the validity of the assumptions imposed by Black-Scholes, then it makes little sense to use a quantity which depends on the parameterization of that model in forecasting a parameter of a different model. If, for example, the true model reproduces the volatility smile, then the volatility at which one arrives by inverting the Black-Scholes formula will be affected by the higher moments of the stock price distribution under the true model.

Therefore, we try to take advantage of the information that implied volatility undoubtedly conveys inasmuch as it is simply a different way of quoting the price of an option. By ensuring comparability between different stocks, this method of quotation is particularly convenient, and making use of it for pricing CDS is of a distinctly different quality compared to the attempt to derive volatility forecasts from it.

To determine what exactly drives implied volatility is well beyond the scope of this book. An increase in implied volatility may reflect market expectations of future volatility – to what extent may be unquantifiable even if the stochastic process for the stock price is known with certainty. Yet it could also result from changes in the higher moments of the stock price distribution, liquidity, model risk, etc. as perceived by the market. Regardless of what causes changes in option-implied volatility may be, this information is potentially useful for the pricing of a CDS.

Despite the considerable uncertainty about what exactly is measured by implied volatility, it seems obvious from the above-mentioned possible explanations – and the apparent lack of alternatives implying the opposite – that an increase is bad news for creditors and should coincide with higher CDS premia. In terms of Merton's model of credit spreads, an increase in option-implied volatility that prices expectations of greater future volatility should be thought of as increasing the probability of a shorter first passage time to default. Anecdotal evidence from financial market coverage in the daily press supports this prediction, for credit spreads and implied volatility have been observed to exhibit sharp increases during times of market turmoil.

5.4 Estimation and Results

The following panel regression is estimated:

$$\begin{aligned}
CDS_t^i = \ &constant + \beta_1\, level_t + \beta_2\, slope_t + \beta_3\, liquidity_t \\
&+ \beta_4\, interest\ coverage_t^i + \beta_5\, leverage_t^i + \beta_6\, profitability_t^i \\
&+ \beta_7\, rating_t^i + \beta_8\, historical\ volatility_{t-1}^i \\
&+ \beta_9\, implied\ volatility_{t-1}^i + \beta_{10}^i firmdummy^i \\
&+ \beta_{11,t}\, timedummy_t + error\ term\,,
\end{aligned}$$

in which CDS_t^i denotes the CDS premium for sample reference entity i at date t.

Note that the formula above represents the most comprehensive model – the other specifications described below contain fewer variables – and that the time dummies are of monthly frequency. The results of the regressions can be found in the appendix. In table C.1 the first column (labeled M0) shows the result of a regression which only uses the general variables interest rate, level, and liquidity along with a constant term. To preserve space, this is the only case reported in which no fixed effects were used to control for variation in the cross section of firms and in the time series. Tables C.1 - C.4 correspond to the four scenarios of regressing CDS premia over the general variables only, general and rating variables, general and accounting variables, and all three types of variables, respectively. Each scenario is subdivided in four specifications which use no volatility information at all (first column), historical volatility as the only volatility measure (second column), implied volatility but not historical volatility (third column) and, finally, historical and implied volatility (fourth column). The dummy variables for the different firms and months are not reported.

A number of conclusions can be drawn from tables C.1 - C.4. As it turns out, option-implied volatility is highly statistically significant in all scenarios under consideration. It also has a pronounced effect on the explanatory power of the regression. The increase in R^2 as compared to the specifications that do not use any volatility information amounts to 5 - 10 percentage points, depending on whether rating information enters the regression. The coefficients take the expected sign and reflect the substantial influence on CDS premia. To gain an understanding of the quantitative effect of option-implied volatility, consider the column labeled M3 in table C.1, for example. An increase in implied volatility of one percentage point in this case will lead to an increase in the estimated CDS premium of 6.43 basis points. In the most general specification (columns M16 in table C.4), this figure is reduced to about four basis points, which is still a sizable magnitude.

Confirming the results of Campbell and Taksler, we find that the coefficients of historical volatility are also statistically significant and have the expected sign. The addition of this information to the set of explanatory variables improves the explanatory power of the regression as well, but to a somewhat lower degree. When used as the only volatility measure, R^2 is raised by 3 - 7 percentage points. The maximum of the estimated effect on CDS premia amounts to 4.58 basis points (M2) and is reduced when historical volatility is used in combination with option-implied volatility.

A comparison of the results from the regressions that use both, historical and implied volatility with those obtained by adding only one of these variables to the regressions suggests the higher relative efficacy of implied volatility in explaining CDS premia. In these cases, the magnitude of the coefficients of historical volatility is about one half of those of implied volatility. In addition, the increase in R^2 is higher when implied volatility is added to a regression already using historical volatility than vice versa.

The use of both the volatility measures together in a regression could potentially cause multi-collinearity since a certain degree of correlation can be expected between these variables. Indeed, the Pearson correlation coefficient for the two volatilities over the whole sample is about 0.75. This is by far the highest correlation between any of the independent variables as can be seen in table 5.2. However, the fact that both variables retain statistical significance when used together along with the improvement of the explanatory power achieved by using both variables indicates that multi-collinearity is probably not an issue. This view is supported by the stability of the coefficient estimates of the other variables, in particular of the important group of rating variables, in the four specifications of each scenario. The changes that can be observed in the liquidity proxy are irrelevant due to the fact that it is not possible to pin down the sign of this variable reliably. The switching sign of the intercept hints at a possibly non-linear relation between implied volatility and CDS premia.

Table 5.2. Pearson Correlation Coefficients

	Implied Volatility	Historical Volatility	Gov5y	Liquidity
Implied Volatility	1.00000	0.75020	0.06681	0.01059
Historical Volatility		1.00000	0.07693	0.02628
Gov5y			1.00000	-0.06266
Liquidity				1.00000

Several interesting results concerning the role of ratings are worth noting. Clearly, ratings are enormously important in explaining CDS premia. The increase in R^2 is the highest for this group of variables. For example, regressing the CDS premia over the general and rating variables (column M5) yields an R^2 which is more than 13 percentage points higher than in the simplest case (column M1). Interestingly, however, this increase is almost matched by the one resulting from adding both volatilities (column M4).

With one (statistically not significant) exception in rating class A, the signs of the coefficients are as expected, and the monotonicity from higher to lower credit quality holds throughout all rating classes. Recall that the effect of ratings is captured relative to the highest rating in the sample, AA.

The difference between the coefficients for a BBB-rating and that for non-investment grade ratings is striking, and so is the statistical significance of these results. This finding can partly be attributed to the dramatic surge of the average cumulative default probability for the respective letter ratings. Over a five year horizon, which corresponds to the maturity of the CDS in the sample, Moody's (2001) [59] reports default probabilities that are almost ten times higher for BB-rated issuers than for those with a BBB rating (11.23 % vs. 1.82%.) If a high probability of default is to be associated with low recovery rates as some authors suggest (e.g. Bakshi et. al (2001) [6]) then this will contribute to an explanation of the discrepancy of the respective spreads.

In addition to this quantifiable cause, the extraordinary increase may indicate the existence of market segmentation or market frictions caused by institutional regulations. The two ratings constitute the threshold between investment grade and non-investment grade credit quality. As some investors are prohibited from investing in junk bonds, they are forced out of their bond-holdings when a downgrade of the debt occurs. The number of market participants investing in high-yield debt is therefore reduced. As a consequence, information quality in this market segment may suffer from this mechanism and liquidity risks may arise. It is plausible that the high coefficients are to a certain extent an outcome of these micro-economic effects.

The performance of the financial ratios in explaining CDS premia as measured by the contribution to R^2 is on the whole disappointing. Nevertheless, looking at the different ratios in greater detail, several noteworthy insights can be gained. The worst performing variable is the proxy for interest coverage. In all scenarios, although at very small values, the coefficient takes the wrong sign. Results from Blume et al. (1998) [13] suggest that changes in interest coverage are much more informative when its level is low. Possibly, breaking the variables into groups rather than measuring them continuously could help to take into account the potential for a non-linear relationship. Campbell and Taksler have followed this advice in their study but to little avail as they, too, are confronted with statistically insignificant estimates and wrong signs.

It is interesting that the profitability proxy is in line with a priori expectations as long as ratings are not used in the regressions (cf. table C.3). The inclusion of ratings in table C.4 leads to a reversal of the estimated sign. In contrast, there seems to be no equivalent effect on the coefficients of the leverage proxy as it keeps the correct sign while retaining statistical significance throughout. Note, however, that the point estimates which remained unaffected by the inclusion of the volatility measures are approximately halved and the t-values are reduced.

This comparison suggests that the information content of the profitability ratio is captured more efficiently by ratings. Contrary to profitability which tends to fluctuate quite drastically, leverage is usually a more slowly moving characteristic. The fact that balance sheet information is purely backward looking therefore puts it at less of a disadvantage with respect to the latter variable.

As for the variables that enter all of the regressions, the liquidity proxy fails to add any explanatory power. Figure 5.1 shows that following the terroristic acts in September the spread between the swap and the government curve exhibits a dramatic surge which resolves into a period of increased volatility. Nevertheless, the coefficients are statistically insignificant.

In contrast, the interest rate information does in part explain CDS premia. With few exceptions, statistical significance of the coefficients is given. Their signs, which are negative in all specifications, confirm earlier findings and support the conjecture about the interrelation between risk-free interest rates and credit spreads.

5.5 Robustness Checks

We investigate the robustness of the results presented in section 5.4 with respect to two aspects. First, we examine whether the results are sensitive to the length of the period used to estimate historical volatility. To this end, we use the annualized standard deviation of daily

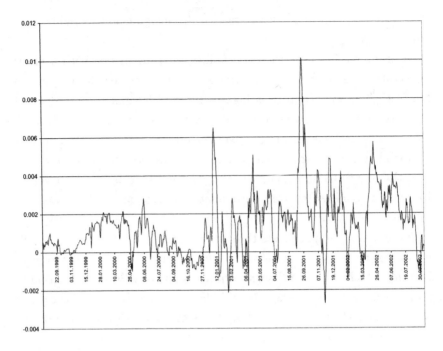

Fig. 5.1. Spread between 3 month swap rate and 3 month T-bill yield

equity returns calculated over 90 and 360 calendar days, respectively. Second, we address the question whether the exact definition of historical volatility is important. In particular, we examine the effect of demeaning the returns in the calculation of the standard deviation by using average squared returns. We also adjust for autocorrelation of returns using the method proposed by French et al. [36]. Finally, we explore average absolute returns as an alternative measure of historical volatility. Thus, in total, we consider five additional measures of volatility. See the appendix for the exact formulae. We re-run the regressions specified in the columns labeled M14 and M16 of table C.4.

The first two columns of tables C.5 and C.6 present evidence that the choice of the horizon used for estimating volatility does not alter the results fundamentally. All estimates are remarkably stable, particularly that of implied volatility. However, it is obviously possible to increase the explanatory power of the regression through a clever choice of the time horizon, as for both periods R^2 is slightly higher than for 180 calendar days.

Similar results hold for the volatilities computed alternatively. To keep the values of the coefficients as comparable as possible, these measures were scaled by the same factor as the standard deviation ($100\sqrt{250}$). As can be seen in the last three columns of tables C.5 and C.6, all estimated coefficients remain virtually unchanged, and R^2 increases slightly.

The analysis of the impact of changing the length of the estimation window is not conducted for the three alternative volatility measures. Since correlations with standard deviation estimated over the respective horizon are well over 90% in all cases, the robustness check performed with the estimates of standard deviation seems sufficient.

5.6 Conclusion

This chapter has provided evidence on the influence of option-implied volatility on CDS premia. To the best of our knowledge, it is the first study to have used this information at the level of individual borrowers in a credit risk context. It is also one of the very few studies that investigate CDS premia as opposed to bond spreads.

The main results from the panel regressions performed on data of CDS quotes referenced to international industrial firms from 1999 to mid-2002 indicate that while both variables are useful in explaining CDS premia, option-implied volatility has a stronger effect than historical volatility. Robustness checks show that the results do not depend on the exact definition of historical volatility or the time span over which it is estimated. Both quantities remain relevant in the presence of credit ratings, which contribute an equal amount of explanatory power. The contribution of term structure information is in line with earlier research. These results are robust to the inclusion of fixed effects for each reference entity and the correction for time series variation.

Regarding topics for future research, the results have documented the need for a more thorough theoretical modeling of the link between volatility and CDS premia. Furthermore, the differences between CDS premia and bond spreads and the interrelation of historical and implied volatility could be investigated both theoretically and empirically. Finally, despite the fact that the robustness checks with respect to several measures of volatility have confirmed the basic results of this study, they have nevertheless documented that it is important to address the issue of measuring volatility very carefully.

6
Conclusion

In the research at hand, we have investigated models of default risk. All empirical work is within the context of reduced-form models. From an academic view, it is regrettable that in these models there is no economic intuition for the default event. However, we have made our choice only after demonstrating that under a strict criterion, practically useful implementations of firm value models do not possess this feature, either. Chapter 3 has presented the necessary theoretical groundwork for the empirical studies in chapters 4 and 5.

The empirical work that we have carried out in chapter 4 on the class of ATSM has made abundantly clear the difficulties to reliably estimate intensity-based models for default risk. In particular, this is the case if one goes beyond the popular class of *completely* ATSM. Despite the impossibility to exactly identify potential sources of errors, their relative importance, or possible interdependences between them, the results suggest that data constraints, numerical complexity, and model misspecification must be addressed.

Data constraints have posed severe problems for empirical studies in the field. We have tried to overcome such limitations by using OTC data from REUTERS and including only firms in the sample for which we were able to estimate a firm-specific term structure. This feature has contributed to the data quality because errors in the records were identified and corrected during the process of estimation. Despite the high demands we make on the data we are able to maintain a remarkable breadth of firms in the sample. The resulting data set is unique in the literature. Our study is the first to work with firm-specific term structures, let alone that for many borrowers this data is available in two denominations.

Nevertheless, our study is not unaffected by data issues, namely because the length of the time series for the individual firms is relatively short. This fact may be responsible in part for the difficulties to obtain statistically significant estimates for the diffusion means of the models. It is questionable, though, whether longer observation periods can indeed be a cure for these problems. We have explained in detail that due to changing firm characteristics, long(er) time series may not be useful at all when estimating the models presented.

The estimation of the models requires numerically highly intensive methods. It is well known that the objective functions possess a vast number of local maxima. We have tried to address this problem by a careful selection of starting values but there is of course no guarantee that the global maxima were reached in the course of the optimizations. On top of this, the parameter combination selected by the optimization routine have occasionally caused conflicts with the machine accuracy. Hence, some parameter constellations may have been rejected solely due to numerical inaccuracies.

Probably more important than these rather technical issues is the potential misspecification of the models, at least in the base case of estimating completely ATSM for defaultable rates. In the other cases, its impact may be equal in absolute terms but as we have seen, the numerical problems increase when the models become more complicated (e.g. essentially ATSM). We have also seen that misspecification is more likely when using spreads simply due to the smaller absolute values of the data. As we have repeatedly pointed out, it is particularly worrying that the in- and out-of-sample errors resulting from the models are in many instances of the same order as those stemming from a simple random walk.

All is not lost, however, as these errors are very small and perhaps smaller than one might have expected given the turbulent time covered by the sample. We have also arrived at a number of interesting results pertaining to the relative efficacy of different modeling approaches and specifications within them, which data to use, what degree of homogeneity can be expected across industries and/or currencies, etc. While some of these insights remain somewhat vague, they may well provide a sensible basis for further research. The examination of alternative models will certainly be an important direction for future work. In the literature on risk-free interest rates, which provide a better playing ground because of better data availability, essentially ATSM have not been the only possibility newly introduced to adequately describe the term structure. Other suggestions include quadratic term structure models and regime switching models.

The challenges for policy making and standard setting that arise from these results are enormous. Keeping in mind that we have obtained our results under substantially simplified model setups, regulations that require a more detailed level of modeling can easily appear unrealistic. How should one be able to extract the components of default risk separately when the dynamics of the spread as a whole cannot be recovered with certainty? How should portfolios of untraded loans (which constitute the overwhelming part of all corporate debt) be treated when there is already so much uncertainty attached to debt for which there is market data?

Clearly, future regulations should aim at increasing the liquidity and transparency of debt markets. This stipulation may appear as a commonplace. Nonetheless, it is entirely justified because the existing regulation has arguably prompted many of the recent developments in credit markets that have had exactly the opposite effect. Particulary popular is the so-called regulatory capital arbitrage. For example, innovative financial instruments such as collateral debt obligations exploit differences in the equity requirements imposed on debt simply by repackaging the components. This complicates an assessment of default risk inherent in such instruments and provides an example of how the existing rules can reduce market transparency. This is all the more worrying because at the same time issuers try to gain better access to

6 Conclusion

capital markets with more complex instruments. While plain instruments such as straight bonds tend to lose market share, bonds with special features such as rating triggered mark-ups on the coupon are issued more frequently. The fragmentation of the market in this respect complicates the calibration of default risk models.

Due to their simple structure, credit default swaps may be a potential data source for this task. In chapter 5, we have presented the results of one of the very first studies using these data. In contrast to the highly ambitious model structures used in the context of ATSM, we have used a much simpler setup. On the one hand, the CDS market is not yet mature enough to provide the sufficient data for the estimation of more complicated models. It is therefore far from being an alternative to bond markets when it comes to the calibration of default risk models. On the other hand, however, there has been an increasing interest in economically insightful results of the kind we were able to find using data from the market for CDS. This interest is well-documented by studies conducted with bond market data that have made a virtue of necessity in a very similar way. We are able to present new evidence on the connection between various measures of equity volatility and CDS premia. The result that implied volatility has a larger impact in our sample than historical volatility is especially noteworthy.

Our experiences from chapter 4 make it seem advisable to continue research along these lines. Unlike the equity and equity derivative markets, there appears to be need of more fundamental research on credit markets. This is not to say that there is no promise in continuing to explore term structure models with defaultable bond data. But there is evidence that at present, the goals of controlling credit risks and ensuring the stability of lending institutions may be better served by simpler models.

A

Calculation of Volatility Proxies

$$\text{Historical Volatility} = 100\sqrt{250}\sqrt{\frac{1}{n}\sum_{i=t-n}^{t-1}\left(r_i - \frac{1}{n}\sum_{i=t-n}^{t-1}r_i\right)^2}$$

$$\text{SqdReturns} = 100\sqrt{250}\sqrt{\frac{1}{n}\sum_{i=t-n}^{t-1}r_i^2}$$

$$\text{AbsReturns} = 100\sqrt{250}\frac{1}{n}\sum_{i=t-n}^{t-1}|r_i|$$

$$\text{AdjVol} = 100\sqrt{250}\sqrt{\frac{1}{n}\sum_{i=t-n}^{t-1}r_i^2 - 2\sum_{i=t-n}^{t-2}r_i r_{i+1}}$$

In the formulae, n denotes the number of returns within the respective period used for estimation (90, 180, or 360 calendar days.) Scaling standard deviation by $100\sqrt{250}$ expresses volatility in annualized percentage points. The scaling is applied to the other measures for comparability of the coefficient estimates.

B

Tables for Chapter 4

This appendix contains the estimation results for Duffee's model. Standard errors are reported in parentheses. Table B.4 presents parameter estimates for the Essentially ATSM estimated for EUR data of Deutsche Bank.

118 B Tables for Chapter 4

Table B.1. Parameter Estimates and Standard Errors (in Parentheses) for Duffee's Model (Financial Firms / EUR Data).

	α	θ	κ	λ	σ	σ_{m1}	σ_{m2}	σ_{m3}	σ_{m4}	β_1	β_2	LogL
Abbey Ntl	0 (5.2e-05)	0.013856 (0.30896)	0.029675 (0.66170)	-0.462352 (0.66236)	0.182951 (0.00180)	0.000381 (1.9e-05)	0 (3e-06)	0.000453 (3.6e-05)	0.000832 (5.2e-05)	-0.012323 (0.00285)	-0.100379 (0.00186)	-17497.8
ABN Amro	0 (0.00015)	1e-05 (0.00917)	0.003223 (0.99694)	-0.091959 (0.99666)	0.085142 (0.00976)	0.000629 (2.1e-05)	0.00013 (1e-06)	0.000532 (4.6e-05)	0.001069 (9.7e-05)	-0.305310 (0.00107)	0.464457 (0.00352)	-10175.9
Commerzbank	0.002695 (5.7e-05)	0.000306 (0.00052)	0.354565 (0.60788)	-0.557694 (0.60716)	0.130274 (0.00377)	0.000927 (3e-05)	0.000205 (6e-06)	0.000138 (6e-06)	0.000427 (1.7e-05)	-0.081477 (0.00267)	-0.144695 (0.00232)	-18775.4
Deutsche Bank	0.003814 (2.2e-05)	0.0018 (0.00037)	0.250963 (0.05114)	-0.370857 (0.05108)	0.074566 (0.0072)	0.000639 (1.4e-05)	0.000171 (2e-06)	0.00017 (1e-06)	0.000352 (4e-06)	-0.099535 (0.00050)	-0.08601 (0.00023)	-21929.9
Dresdner Bank	0.001949 (7.6e-05)	0.000759 (0.00229)	0.310507 (0.93868)	-0.559717 (0.93812)	0.027997 (0.01642)	0.000514 (2.1e-05)	0.00034 (2.6e-05)	0.000308 (2.1e-05)	0.000516 (2.5e-05)	0.005512 (0.00274)	-0.037291 (0.00247)	-16134.2
Goldman Sachs	0 (0.00019)	0.000897 (0.05051)	0.115919 (6.5314)	-0.320043 (6.53136)	0.08732 (0.00377)	0.000833 (4.2e-05)	0.000285 (1.1e-05)	0.000113 (3.7e-05)	0.000491 (2.2e-05)	-0.027009 (0.00478)	0.07028 (0.00687)	-12579.9
HypoVereinsbank	0.001196 (8e-05)	1e-05 (5.4e-05)	0.083935 (0.18402)	-0.628168 (0.18440)	0.091981 (0.00323)	0.000818 (2.7e-05)	0.000443 (2.5e-05)	0.000263 (2.7e-05)	0.000991 (7.1e-05)	0.026486 (0.00336)	-0.008248 (0.00298)	-19715.0
ING	0 (0.00019)	0.0025 (0.00497)	0.130607 (0.26061)	-0.317666 (0.26073)	0.031816 (0.00852)	0.00193 (0.00016)	0.000727 (7.2e-05)	0.000282 (3.1e-05)	0.000667 (4.3e-05)	0.025822 (0.00871)	0.031298 (0.00657)	-13766.8
Lehman Bros.	0.004791 (0.00024)	0.00866 (0.04725)	1.24362 (6.7924)	-0.903675 (6.7881)	0.044202 (0.38023)	0.000555 (1.9e-05)	0.000423 (1.9e-05)	0.000777 (4.7e-05)	0.001827 (0.00013)	-0.267318 (0.00518)	-0.412776 (0.00258)	-10142.8
Merrill Lynch	0.015462 (0.00020)	0.01117 (0.93487)	1.20779 (101.069)	-0.685236 (101.07)	0.105446 (0.14941)	0.000397 (2e-06)	0.000370 (6e-06)	0.000373 (6e-06)	0.00093 (2.4e-05)	-0.60673 (0.00076)	-0.498033 (0.00047)	-5759.8
Rabobank	0.00107 (0.00012)	0.00657 (44.8597)	3.5e-05 (0.24068)	-0.306847 (0.24133)	0.083275 (0.00178)	0.000173 (4e-05)	9.1e-05 (3.2e-05)	1e-05 (3.5e-05)	0.000118 (2.6e-05)	0.045933 (0.00426)	-0.155961 (0.00466)	-14383.7

Table B.2. Parameter Estimates and Standard Errors (in Parentheses) for Duffee's Model (Industrial Firms / EUR Data).

	α	θ	κ	λ	σ	σ_{m1}	σ_{m2}	σ_{m3}	σ_{m4}	β_1	β_2	LogL
DaimlerChrysler	0 (0.00051)	0.011725 (0.41926)	0.072426 (2.58985)	-0.226953 (2.58945)	0.058109 (0.01971)	0.001193 (4.8e-05)	0.00023 (1.4e-05)	0.000765 (6.4e-05)	0.001787 (0.00016)	-0.09192 (0.01238)	0.073161 (0.01432)	-8391.2
Endesa	0.010014 (0.00068)	0.005074 (0.21516)	1.03621 (43.9403)	-0.826221 (43.8953)	0.035721 (2.24129)	0.00068 (2.0e-05)	0.000771 (7.6e-05)	0.000876 (8.8e-05)	0.000742 (2.3e-05)	-0.263303 (0.00662)	-0.459861 (0.00599)	-4299.8
Fiat	0 (0.00039)	0.016491 (0.39165)	0.049593 (1.17795)	-0.345075 (1.17961)	0.152302 (0.00384)	0.000898 (8.2e-05)	0.000271 (1.9e-05)	0 (0.00023)	0.000486 (2.3e-05)	-0.007673 (0.01075)	0.050070 (0.01048)	-8915.0
Ford	0.028525 (0.00018)	0.008512 (1.12289)	0.078821 (10.4024)	-0.421143 (10.3899)	0.211429 (0.01121)	0.00085 (1.3e-05)	0.000658 (1.9e-05)	1.5e-05 (0)	0.000559 (8.0e-06)	-0.625962 (4.7e-05)	-0.41392 (2.4e-05)	-5702.0
General Electric	0.003918 (0.00010)	1.9e-05 (1.2e-05)	0.674363 (0.41267)	-0.963274 (0.41240)	0.071134 (0.00253)	0.000373 (1.6e-05)	0.000171 (1.2e-05)	3.2e-05 (3.0e-06)	0.00029 (2.2e-05)	-0.056441 (0.00352)	-0.131941 (0.00301)	-9957.2
General Motors	0 (0.00046)	2.1e-05 (3.1e-05)	0.676707 (0.10957)	-0.999061 (0.10951)	0.121958 (0.00397)	0.00118 (8.7e-05)	0.000486 (2.8e-05)	0 (0.00051)	0.00074 (4.1e-05)	-0.140834 (0.01110)	0.174364 (0.01395)	-7741.9
KPN	0 (0.00133)	0.005034 (0.22326)	0.092971 (4.11897)	-0.276487 (4.11775)	0.105706 (0.00798)	0.001909 (0.00023)	0.000797 (9.5e-05)	0 (0.00104)	0.001242 (0.00014)	-0.365274 (0.03199)	0.089660 (0.03915)	-4242.2
McDonalds	0.001983 (8.5e-05)	0.009337 (0.09764)	0.049443 (0.51711)	-0.287021 (0.51707)	0.138076 (0.00612)	0.000988 (5.8e-05)	0.000339 (2.0e-05)	0.000113 (2.0e-06)	0.000448 (1.6e-05)	-0.094854 (0.00255)	-0.030813 (0.00175)	-9950.8
Parmalat	0.001713 (0.00104)	0.000139 (0.01543)	0.071535 (7.91448)	-0.406823 (7.91502)	0.132975 (0.00602)	0.001427 (0.00011)	0.000698 (7.2e-05)	0 (0.00071)	0.00137 (0.00016)	-0.145177 (0.02501)	0.189844 (0.03572)	-6082.8
Philip Morris	0.008371 (3.5e-05)	1.2e-05 (1.9e-05)	0.465934 (0.54235)	-0.804844 (0.54235)	0.126552 (0.00163)	0.001471 (2.9e-05)	0.000522 (1.7e-05)	0 (1.0e-06)	0.000902 (3.7e-05)	-0.216113 (0.00039)	-0.045467 (0.00035)	-18090.3
Total	0.006349 (8.1e-05)	0.000102 (5e-05)	0.662305 (0.31607)	-1.068140 (0.31579)	0.078040 (0.00710)	0.000847 (5.8e-05)	0.000267 (1.5e-05)	9.5e-05 (1.1e-05)	0.000183 (1.2e-05)	-0.062667 (0.00253)	-0.145881 (0.00243)	-7023.2
Toyota	0.010291 (3.4e-05)	7.5e-05 (0.00083)	0.028249 (0.27633)	-0.244445 (0.27635)	0.057448 (0.01659)	0.000386 (1.0e-05)	0.00017 (1.0e-06)	0.000689 (2.8e-05)	0.001508 (7.2e-05)	-0.278813 (0.00072)	-0.336099 (0.00055)	-16265.7

Table B.3. Parameter Estimates and Standard Errors (in Parentheses) for Duffee's Model (USD Data).

	α	θ	κ	λ	σ	σ_{m1}	σ_{m2}	σ_{m3}	σ_{m4}	β_1	β_2	LogL
Abbey Ntl	0	0.018762	0.08507	-0.540931	0.208211	0.000925	0	0.001155	0.002490	0.071392	-0.370641	-14746.8
	(0.00021)	(0.32950)	(1.49452)	(1.49691)	(0.00569)	(6.3e-05)	(0.00041)	(0.00037)	(0.00067)	(0.00532)	(0.00857)	
ABN Amro	0.021474	0.005313	1.24568	-1.12479	0.063813	0.001689	0.000216	0.000822	0.001596	-0.681231	-0.501403	-14035.0
	(7.5e-05)	(0.01619)	(3.79817)	(3.79802)	(0.01291)	(4.0e-05)	(1.0e-06)	(2.0e-05)	(3.3e-05)	(0.00144)	(0.00128)	
Bank of America	0.002491	0.009118	1.0322	-0.924621	0.198769	0.000755	0	0.000436	0.000860	0.075510	-0.725499	-6009.1
	(0.00040)	(0.01379)	(1.55831)	(1.55854)	(0.02292)	(5.7e-05)	(9.9e-05)	(6.4e-05)	(8.6e-05)	(0.00176)	(0.00303)	
Citibank	0.004563	0.00629	1.37407	-1.20779	0.094617	0.001476	5.5e-05	0.000302	0.000731	-0.195524	-0.483124	-12300.2
	(0.00018)	(0.00496)	(1.08653)	(1.08699)	(0.01159)	(0.00010)	(5.0e-06)	(1.5e-05)	(1.3e-05)	(0.00172)	(0.00289)	
Deutsche Bank	0.002214	0.004642	0.372056	-0.674072	0.197234	0.000731	2.8e-05	0.000512	0.001827	0.168306	-0.731443	-12652.3
	(0.00022)	(0.00231)	(0.18520)	(0.18412)	(0.00278)	(3.1e-05)	(5.4e-05)	(0.00011)	(0.00034)	(0.00509)	(0.01036)	
DaimlerChrysler	0	0.000748	0.23198	-0.488034	0.142088	0.001421	0.000586	0	0.000696	-0.178624	-0.019558	-9622.0
	(0.00031)	(0.00660)	(2.04978)	(2.04983)	(0.00174)	(0.00012)	(5.4e-05)	(0.00102)	(3.6e-05)	(0.00643)	(0.00999)	
Dresdner Bank	0	0.015041	0.295194	-0.332937	0.069433	0.001562	0.000298	0.000903	0.002016	-0.005593	-0.362542	-14694.4
	(0.00024)	(0.44096)	(8.65388)	(8.65064)	(0.01682)	(0.00010)	(2.1e-05)	(9.6e-05)	(0.000223)	(0.00530)	(0.00745)	
Ford	0.008745	0.001449	0.236409	-0.564898	0.149358	0.001593	0.000455	0	0.000358	-0.148390	-0.069019	-18435.5
	(0.00012)	(0.00224)	(0.36594)	(0.36616)	(0.00097)	(9.0e-05)	(2.2e-05)	(0.00050)	(9.0e-06)	(0.00371)	(0.00317)	
General Electric	0	0.000822	0.356068	-0.611883	0.144642	0.001223	0.000374	0	0.000253	-0.099467	0.028591	-19488.8
	(0.00014)	(0.00430)	(1.86667)	(1.8672)	(0.00142)	(9.9e-05)	(2.8e-05)	(0.00070)	(9.0e-06)	(0.00434)	(0.00541)	

B Tables for Chapter 4   121

Table B.3 (cont.). Parameter Estimates and Standard Errors (in Parentheses) for Duffee's Model (USD Data).

General Motors	0.015598	0.010238	0.027075	-0.383304	0.146151	0.001698	0.000547	0	0.000405	-0.300255	0.043263	-17708.7
	(7.8e-05)	(0.29752)	(0.78689)	(0.00100)	(0.00011)	(4.1e-05)	(5.0e-06)		(1.0e-05)	(0.00246)	(0.00127)	
Goldman Sachs	0.018483	0.013309	1.15893	-0.903972	0.079638	0.002747	0.000321	0.000864	0.001089	-0.871893	-0.466789	-7502.2
	(0.00026)	(0.07409)	(6.45924)	(6.45896)	(0.03180)	(6.3e-05)	(1.0e-06)	(1.0e-05)	(9.0e-06)	(0.00195)	(0.00141)	
HypoVereinsbank	0.020242	1.0e-05	0.185833	-1.11496	0.188396	0.000612	0.000276	0	0.000391	-0.341458	-0.092232	-8691.8
	(3.2e-05)	(0.00018)	(3.22721)	(3.22766)	(0.00305)	(3.0e-05)	(1.9e-05)		(2.6e-05)	(0.00031)	(0.00019)	
ING	0	0.007112	1.27947	-1.18146	0.102323	0.000420	0	0.000782	0.001626	-0.233834	-0.400355	-13221.3
		(0.00013)	(0.00796)	(1.43349)	(1.43314)	(0.01037)	(2.0e-05)	(5.2e-05)	(7.9e-05)	(0.00012)	(0.00190)	(0.00250)
Lehman Bros.	0.024479	0.005099	1.21097	-1.12967	0.124993	0.001081	0.000161	0.000353	0.000687	-0.486405	-0.407098	-5293.9
	(8.3e-05)	(0.00015)	(0.03674)	(0.03621)	(0.00951)	(3.1e-05)	(0)	(3.0e-06)	(8.0e-06)	(0.00071)	(0.00054)	
Merrill Lynch	0.011178	0.001272	0.475836	-0.822054	0.157381	0.001633	0.000501	0	0.000370	-0.149756	-0.189899	-18523.1
	(0.00012)	(0.00128)	(0.47839)	(0.47843)	(0.00112)	(0.00011)	(2.6e-05)	(0.00024)	(8.0e-06)	(0.00409)	(0.00351)	
Rabobank	0.007954	0.007453	0.561889	-0.51321	0.053669	0.001200	0.000109	0.000368	0.000639	-0.387492	-0.443503	-12011.7
	(0.00013)	(0.03403)	(2.56608)	(2.56573)	(0.01731)	(8.6e-05)	(2.1e-05)	(1.4e-05)	(1.4e-05)	(0.00281)	(0.00557)	
Toyota	0	0.019703	0.356426	-0.222933	0.051863	0.000998	0.000271	0.00084	0.001701	-0.247162	-0.231133	-10198.6
		(0.00033)	(1.48974)	(26.9551)	(26.9529)	(0.02937)	(5.0e-05)	(2.1e-05)	(9.9e-05)	(0.00014)	(0.00479)	(0.00760)

Table B.4. Parameter Estimates for Essentially ATSM for EUR Data of Deutsche Bank.

	$A_0(3)$	$A_1(3)$	$A_2(3)$	$A_3(3)$
δ_0	0.0191	0.0123	0.0597	0.0351
δ_1	-0.0012	0.0000	0.0002	0.0000
δ_2	0.0137	0.0006	0.0010	0.0111
δ_3	0.0079	0.0034	0.0003	0.0000
$\lambda_{1(1)}$	-0.8543	-0.0751	-5.1501	-2.3464
$\lambda_{1(2)}$	0.0357	-0.5086	0.5764	24.9015
$\lambda_{1(3)}$	-3.2265	1.7269	1.5921	1.6069
κ_{11}	0.0568	0.0342	5.4154	2.7503
κ_{12}			-12.1195	-11.0486
κ_{13}				-0.0280
κ_{21}	0.7057	-0.0001	-0.0162	-0.6293
κ_{22}	-0.3530	-1.3941	0.0392	9.1784
κ_{23}		-2.3170		-0.7051
κ_{31}	-0.9064	-1.4536	0.0000	-3.0882
κ_{32}	1.5833	-0.6753	0.0000	-0.0389
κ_{33}	-0.4100	1.9014	-2.9647	5.7931
β_{12}		1.2520		
β_{13}		0.2858	0.0007	
β_{23}			0.0269	
$\lambda_{2(11)}$	-0.0866			
$\lambda_{2(12)}$	0.1810			
$\lambda_{2(13)}$	0.2114			
$\lambda_{2(21)}$	-0.5839	0.5270		
$\lambda_{2(22)}$	0.5814	1.3483		
$\lambda_{2(23)}$	-0.0895	3.3209		
$\lambda_{2(31)}$	1.1665	0.6860	0.2193	
$\lambda_{2(32)}$	0.6841	0.3778	-2.9445	
$\lambda_{2(33)}$	1.3929	-0.1743	3.1428	
θ_1		7.0758	12.3744	11.6632
θ_2			5.4445	2.5772
θ_3				6.0875
$\chi^2(df)$	854.84	9482.86	9619.47	22000.00
df	14	13	15	17

C

Tables for Chapter 5

In the following tables the results of the regression analyses of chapter 5 are presented. All tables show the estimated coefficients, t-values are reported in parentheses. Tables C.1 - C.4 report the results of the different scenarios using information on volatility, ratings, and accounting ratios in several combinations. All regressions except for the one labeled M0 use fixed effects to correct for time series and cross issuer variation of the CDS spreads.

Tables C.5 and C.6 investigate the effect of using different alternatives in calculating historical volatility. The heading of each column indicates which exact measure has been used in the respective regression.

Table C.1. Results Using General and Volatility Variables

Regression	M0	M1	M2	M3	M4
Historical Volatility			0.0458		0.0282
			(73.58)		(44.11)
Implied Volatility				0.0643	0.0497
				(88.71)	(64.30)
A					
BBB					
BB					
B					
Ebit/NetSales					
Ebit/TotInterest					
LtDebt/TotAssets					
Gov5y	-0.4871	-0.3970	-0.3762	-0.0780	-0.0710
	(-34.01)	(-6.58)	(-6.84)	(-1.47)	(-1.38)
Liquidity	1.0817	0.0666	2.0696	-0.8363	0.6031
	(0.36)	(0.03)	(0.98)	(-0.41)	(0.31)
Intercept	3.5232	2.6256	1.5838	-1.8895	-1.5095
	(47.67)	(7.03)	(4.65)	(-5.69)	(-4.71)
RMSE	1.7024	1.1514	1.0492	1.0110	0.9758
RSQ	0.0419	0.5641	0.6381	0.6640	0.6870

Table C.2. Results Using General, Rating, and Volatility Variables

Regression	M5	M6	M7	M8
Historical Volatility		0.0328		0.0195
		(59.41)		(34.62)
Implied Volatility			0.0497	0.0405
			(78.28)	(59.96)
A	0.0210	0.2310	0.0574	0.1756
	(0.55)	(6.38)	(1.66)	(5.16)
BBB	0.6577	0.7987	0.6065	0.6999
	(12.75)	(16.47)	(13.05)	(15.37)
BB	9.9319	8.8068	8.4906	8.0883
	(93.82)	(87.00)	(87.39)	(84.49)
B	11.9621	10.9551	10.0189	9.7798
	(59.97)	(58.23)	(55.21)	(55.06)
Ebit/NetSales				
Ebit/TotInterest				
LtDebt/TotAssets				
Gov5y	-0.3879	-0.3764	-0.0768	-0.1276
	(-7.74)	(-8.00)	(-1.69)	(-2.88)
Liquidity	1.6314	2.8869	0.7070	1.6249
	(0.85)	(1.60)	(0.41)	(0.96)
Intercept	2.6577	1.7245	-0.8705	-0.7723
	(8.52)	(5.88)	(-3.06)	(-2.77)
RMSE	0.9559	0.8979	0.8614	0.8425
RSQ	0.6997	0.7350	0.7561	0.7667

Table C.3. Results Using General, Accounting, and Volatility Variables

Regression	M9	M10	M11	M12
Historical Volatility		0.0450		0.0278
		(71.86)		(43.35)
Implied Volatility			0.0634	0.0494
			(87.43)	(64.05)
A				
BBB				
BB				
B				
Ebit/NetSales	-0.0147	-0.0052	-0.0057	-0.0018
	(-12.31)	(-4.70)	(-5.43)	(-1.79)
Ebit/TotInterest	0.0001	0.0005	0.0002	0.0004
	(0.64)	(3.15)	(1.62)	(3.10)
LtDebt/TotAssets	0.0330	0.0236	0.0250	0.0209
	(16.89)	(13.15)	(14.48)	(12.53)
Gov5y	-0.4019	-0.3796	-0.0004	-0.0751
	(-6.71)	(-6.92)	(-0.01)	(-1.47)
Liquidity	-0.0630	1.6949	-1.1076	0.2094
	(-0.03)	(0.80)	(-0.55)	(0.11)
Intercept	1.1879	0.4841	-3.0065	-2.5160
	(3.10)	(1.38)	(-8.81)	(-7.62)
RMSE	1.1428	1.0454	1.0067	0.9727
RSQ	0.5707	0.6408	0.6669	0.6890

Table C.4. Results Using All Variables

Regression	M13	M14	M15	M16
Historical Volatility		0.0329		0.0197
		(59.24)		(34.82)
Implied Volatility			0.0497	0.0406
			(78.33)	(60.31)
A	-0.0546	0.1907	0.0060	0.1415
	(-1.40)	(5.18)	(0.17)	(4.10)
BBB	0.5231	0.7454	0.5325	0.6637
	(9.81)	(14.84)	(11.09)	(14.09)
BB	9.7415	8.7758	8.4291	8.0919
	(89.69)	(84.91)	(84.90)	(82.94)
B	11.8115	10.9193	9.9588	9.7645
	(59.09)	(57.95)	(54.82)	(54.94)
Ebit/NetSales	0.0003	0.0057	0.0053	0.0076
	(0.27)	(5.95)	(5.79)	(8.49)
Ebit/TotInterest	0.0000	0.0003	0.0001	0.0003
	(0.06)	(2.34)	(1.02)	(2.31)
LtDebt/TotAssets	0.0183	0.0111	0.0135	0.0101
	(10.87)	(7.00)	(8.86)	(6.76)
Gov5y	-0.3887	-0.3761	-0.0760	-0.1257
	(-7.78)	(-8.01)	(-1.68)	(-2.84)
Liquidity	1.3190	2.4870	0.3191	1.2006
	(0.69)	(1.38)	(0.19)	(0.71)
Intercept	1.7908	1.0987	-1.5997	-1.3929
	(5.58)	(3.64)	(-5.47)	(-4.87)
RMSE	0.9538	0.8962	0.8594	0.8404
RSQ	0.7010	0.7360	0.7573	0.7679

Table C.5. Robustness Check without Implied Volatility

Volatility Measure	StDev90	StDev360	SqdReturns	AbsReturns	AdjVol
Historical Volatility	0.0204	0.0175	0.0284	0.0637	0.0080
	(43.25)	(35.05)	(52.06)	(57.54)	(17.73)
A	0.1061	0.1431	0.2096	0.1169	0.0804
	(2.96)	(3.94)	(5.92)	(3.35)	(2.17)
BBB	0.7217	0.7132	0.8149	0.6796	0.7078
	(14.74)	(14.40)	(16.86)	(14.24)	(14.02)
BB	8.5562	8.6001	8.2402	8.1138	8.8066
	(84.72)	(84.13)	(82.37)	(81.82)	(84.88)
B	10.6567	10.3451	10.1731	9.6737	10.6912
	(56.97)	(54.58)	(55.10)	(52.76)	(55.55)
Ebit/NetSales	0.0002	0.0022	0.0035	0.0066	0.0006
	(0.23)	(2.32)	(3.81)	(7.16)	(0.59)
Ebit/TotInterest	0.0000	0.0002	0.0002	0.0002	0.0001
	(0.10)	(1.50)	(1.78)	(2.00)	(0.56)
LtDebt/TotAssets	0.0094	0.0113	0.0066	0.0067	0.0096
	(6.04)	(7.18)	(4.28)	(4.41)	(6.01)
Gov5y	-0.3536	-0.3707	-0.3783	-0.4013	-0.3925
	(-7.71)	(-7.99)	(-8.37)	(-8.97)	(-8.32)
Liquidity	1.0325	2.1413	2.4580	3.4465	1.6671
	(0.59)	(1.20)	(1.42)	(2.01)	(0.92)
Intercept	1.4089	1.3913	1.4291	1.1814	1.9928
	(4.77)	(4.65)	(4.91)	(4.10)	(6.56)
RMSE	0.8741	0.8842	0.8613	0.8526	0.8992
RSQ	0.7501	0.7443	0.7573	0.7622	0.7355

Table C.6. Robustness Check with Implied Volatility

Volatility Measure	StDev90	StDev360	SqdReturns	AbsReturns	AdjVol
Implied Volatility	0.0395	0.0417	0.0371	0.0349	0.0439
	(57.89)	(66.54)	(56.24)	(50.62)	(69.01)
Historical Volatility	0.0085	0.0118	0.0168	0.0360	0.0021
	(17.32)	(25.24)	(30.29)	(30.24)	(5.05)
A	0.0846	0.1370	0.1609	0.1011	0.0672
	(2.51)	(4.07)	(4.81)	(3.03)	(1.97)
BBB	0.6583	0.6728	0.7287	0.6484	0.6408
	(14.27)	(14.68)	(15.95)	(14.23)	(13.79)
BB	7.8874	7.7726	7.6896	7.6855	7.9124
	(82.31)	(81.48)	(80.92)	(80.86)	(82.12)
B	9.4201	9.1267	9.2024	9.0048	9.2985
	(53.08)	(51.77)	(52.49)	(51.30)	(52.16)
Ebit/NetSales	0.0035	0.0056	0.0055	0.0069	0.0038
	(4.02)	(6.35)	(6.34)	(7.85)	(4.33)
Ebit/TotInterest	0.0001	0.0002	0.0002	0.0002	0.0001
	(0.71)	(1.91)	(1.80)	(1.80)	(0.86)
LtDebt/TotAssets	0.0081	0.0087	0.0062	0.0065	0.0081
	(5.50)	(5.97)	(4.27)	(4.50)	(5.53)
Gov5y	-0.1259	-0.1142	-0.1493	-0.1765	-0.1147
	(-2.90)	(-2.65)	(-3.48)	(-4.11)	(-2.63)
Liquidity	0.4178	0.9991	1.2343	1.8158	0.5596
	(0.15)	(0.61)	(0.75)	(1.11)	(0.34)
Intercept	-0.9494	-1.2905	-0.8890	-0.8585	-0.9934
	(-3.37)	(-4.61)	(-3.20)	(-3.09)	(-3.51)
RMSE	0.8232	0.8181	0.8138	0.8139	0.8275
RSQ	0.7783	0.7811	0.7833	0.7833	0.7760

References

1. Viral D. Acharya and Jennifer N. Carpenter. Callable Defaultable Bonds: Valuation, Hedging, and Optimal Exercise Boundaries. Working Paper, Stern School of Business, NYU, 1999.
2. Edward I. Altman, Brooks Brady, Andrea Resti, and Andrea Sironi. The Link between Default and Recovery Rates: Implications for Credit Risk Models and Procyclicality. Working Paper, Stern School of Business, 2002.
3. Edward I. Altman and D. L. Kao. Rating Drift in High-Yield Bonds. *Journal of Fixed Income*, (2):15–20, 1992.
4. Edward I. Altman and Vellore M. Kishore. Almost Everything You Wanted to Know about Recoveries on Defaulted Bonds. *Financial Analysts Journal*, (6):57–64, 1996.
5. Torben G. Andersen and Jesper Lund. Estimating Continuous-Time Stochastic Volatility Models of the Short-Term Interest Rate. *Journal of Econometrics*, 77:343–377, 1997.
6. Gurdip Bakshi, Dilip Madan, and Frank Zhang. Investigating the Sources of Default Risk: Lessons from Empirically Evaluating Credit Risk Models. Working Paper, University of Maryland and Federal Reserve Board, 2001.
7. David S. Bates. Post-'87 Crash Fears in the S&P 500 Futures Option Market. *Journal of Econometrics*, 94:181–238, 2000.
8. David S. Bates. Empirical Option Pricing: A Retrospection. Working Paper, University of Iowa, 2002.
9. Christoph Benkert. Explaining Credit Default Swap Premia. *Journal of Futures Marktes*, 24(1):71–92, 2004.

10. Andreas Binder, Ruth Fingerlos, Rainer Jankowitsch, Stefan Pichler, and Waltraud Zeipelt. Die Schätzung der österreichischen Zinsstruktur nach dem Verfahren von Svensson. Working Paper, Österreichische Kontrollbank, 1999.
11. Fischer Black and John C. Cox. Valuing Corporate Securities: Some Effects of Bond Indenture Provisions. *Journal of Finance*, 31:351–367, 1976.
12. Fischer Black and Myron Scholes. The Pricing of Options and Corporate Liabilities. *Journal of Political Economy*, 81:637–654, 1973.
13. Marshall E. Blume, Felix Lim, and Craig A. MacKinlay. The Declining Credit Quality of U.S. Corporate Debt. *The Journal of Finance*, 53(4):1389–1413, 1998.
14. Eric Briys and Francois de Varenne. Valuing Risky Fixed Rate Debt: An Extension. *Journal of Financial and Quantitative Analysis*, 32:239–248, 1997.
15. John Campbell, Andrew Lo, and Craig MacKinlay. *The Econometrics of Financial Markets*. Princeton University Press, Princeton, 1. edition, 1997.
16. John Y. Campbell, Martin Lettau, Burton G. Malkiel, and Yexiao Xu. Have Individual Stocks Become More Volatile? An Empirical Explanation of Idiosyncratic Risk. *Journal of Finance*, 56(1):1–46, 2001.
17. John Y. Campbell and Glen Taksler. Equity Volatility and Corporate Bond Yields. Working Paper, NBER No. 8961, forthcoming Journal of Finance, 2003.
18. George Chacko. Risk Premium Factors in the Term Structure of Interest Rates. Working Paper, Harvard University, 1997.
19. Pierre Collin-Dufresne, Robert S. Goldstein, and J. Spencer Martin. The Determinants of Credit Spread Changes. *Journal of Finance*, 56(6):2177–2207, 2001.
20. John C. Cox, Jonathan E. Ingersoll, Jr., and Stephen A. Ross. A Theory of the Term Structure of Interest Rates. *Econometrica*, 53:385–407, 1985.
21. Qiang Dai and Kenneth J. Singleton. Specification Analysis of Affine Term Structure Models. *The Journal of Finance*, 55(5):1943–1978, 2000.
22. Jin-Chuan Duan and Jean-Guy Simonato. Estimating and Testing Exponential-Affine Term Structure Models by Kalman Filter. *Review of Quantitative Finance and Accounting*, 13:111–135, 1999.
23. Jefferson Duarte. Evaluating An Alternative Risk Preference in Affine Term Structure Models. *Review of Financial Studies*, 2003.

24. Gregory R. Duffee. The Relation between Treasury Yields and Corporate Bond Yield Spreads. *Journal of Finance*, 53(6):2225–2241, 1998.
25. Gregory R. Duffee. Estimating the Price of Default Risk. *Review of Financial Studies*, 12(1):197–226, 1999.
26. Gregory R. Duffee. Term Premia and Interest Rate Forecasts in Affine Models. *Journal of Finance*, 57(1):405–443, 2002.
27. Gregory R. Duffee and Richard H. Stanton. Estimation of Dynamic Term Structure Models. Working Paper, Haas School of Business, UC Berkeley, 2001.
28. Darrell Duffie. Credit Swap Valuation. Working Paper, Stanford University, 1998.
29. Darrell Duffie, D. Filipovi, and W. Schachermayer. Affine Processes and Applications in Finance. Working Paper, NBER No. 281, 2002.
30. Darrell Duffie and Ming Huang. Swap Rates and Credit Quality. *Journal of Finance*, 51:921–949, 1996.
31. Darrell Duffie and David Lando. Term Structures of Credit Spreads with Incomplete Accounting Information. *Econometrica*, 69:633–664, 2001.
32. Darrell Duffie and Kenneth Singleton. Modeling Term Structures of Defaultable Bonds. *Review of Financial Studies*, 12(4):687–720, 1999.
33. Young Ho Eom, Jean Helwege, and Jing-zhi Huang. Structural Models of Corporate Bond Pricing: An Empirical Analysis. Working Paper, Yonsei University, 2002.
34. Jan Ericsson and Joel Reneby. Estimating Structural Bond Pricing Models. Working Paper, McGill University and Stockholm School of Economics, 2002.
35. Jan Ericsson and Joel Reneby. The Valuation of Corporate Liabilities: Theory and Tests. Working Paper, McGill University and Stockholm School of Economics, 2002.
36. Kenneth R. French, G. William Schwert, and Robert F. Stambaugh. Expected Stock Returns and Volatility. *Journal of Financial Economics*, 19:3–29, 1987.
37. Jon Frye. Depressing Recoveries. *Risk*, pages 108–111, 2000.
38. Ronald Gallant and George Tauchen. Which Moments to Match. *Econometric Theory*, 12:657–681, 1996.
39. Ronald Gallant and George Tauchen. Efficient Method of Moments. Working Paper, University of North Carolina, 2001.

40. Ronald Gallant and George Tauchen. EMM: A Program for Efficient Method of Moments Estimation. Working Paper, University of North Carolina, 2001.
41. Ronald Gallant and George Tauchen. SNP: A Program for Nonparametric Time Series Analysis. Working Paper, University of North Carolina, 2001.
42. Ronald Gallant and George Tauchen. Simulated Score Methods and n direct nference for Continuous time Models. Working Paper, University of North Carolina, 2002.
43. Ro ert Geske. The aluation of Corporate ia ilities as Compound p tions. *Journal of Financial and Quantitative Analysis*, 12:541–552, 1977.
44. Amit Goyal and Pedro Santa-Clara. Idiosyncratic Risk Matters! Working Paper, UCLA, 2002.
45. Steven L. Heston. A Closed-Form Solution for Options with Stochastic Volatility with Applications to Bond and Currency Options. *Review of Financial Studies*, 6(2):327–343, 1993.
46. Patrick Houweling and Ton Vorst. An Empirical Comparison of Default Swap Models. Working Paper, Erasmus University Rotterdam, 2001.
47. Robert Jarrow, David Lando, and Stuart Turnbull. A Markov Model for the Term Structure of Credit Risk. *Review of Financial Studies*, 10:481–523, 1997.
48. Robert Jarrow and Stuart Turnbull. Pricing Derivatives on Financial Securities Subject to Credit Risk. *Journal of Finance*, 50:53–85, 1995.
49. Alexander Kempf and Marliese Uhrig-Homburg. Liquidity And Its Impact On Bond Prices. *Schmalenbach Business Review*, 52:26–44, 2000.
50. David Lando and Torben M. Skødeberg. Analyzing Rating Transitions and Rating Drift with Continuous Observations. *Journal of Banking and Finance*, 26:423–444, 2002.
51. Francis A. Longstaff and Eduardo S. Schwartz. A Simple Approach to Valuing Risky Fixed and Floating Rate Debt. *Journal of Finance*, 50:789–819, 1995.
52. Dilip Madan and Haluk Unal. Pricing the Risks of Default. *Review of Derivatives Research*, 2:121–160, 1998.
53. Robert C. Merton. On the Pricing of Corporate Debt: The Risk Structure of Interest Rates. *Journal of Finance*, 29:449–470, 1974.
54. Charles R. Nelson and Andrew F. Siegel. Parsimonious Modelling of Yield Curves. *Journal of Business*, 60(4):473–489, 1987.

55. Neil D. Pearson and Tong-Sheng Sun. Exploiting the Conditional Density in Estimating the Term Structure: An Application to the Cox, Ingersoll, and Ross Model. *Journal of Finance*, 49(4):1279–1304, 1994.
56. Riccardo Rebonato. *Interest-Rate Option Models*. John Wiley & Sons, New York, 2. edition, 1998.
57. Jesús Saá-Requejo and Pedro Santa-Clara. Bond Pricing with Default Risk. Working Paper, University of California, Los Angeles, 1999.
58. Rainer Sch"obel. A Note on the Valuation of Risky Corporate Bonds. *OR Spektrum*, 21:35–47, 1999.
59. Moody's Investor Service. Default and Recovery Rates of Corporate Bond Issuers: 2000. Working Paper, 2001.
60. David C. Shimko, Naohiko Tejima, and Donald R. Van Deventer. The Pricing of Risky Debt When Interest Rates are Stochastic. *Journal of Fixed Income*, pages 58–65, September 1993.
61. Elias M. Stein and Jeremy C. Stein. Stock Price Distributions with Stochastic Volatility: An Analytic Approach. *Review of Financial Studies*, 4(4):727–752, 1991.
62. Lars E. O. Svensson. Estimating and Interpreting Forward Interest Rates. Working Paper, National Bureau of Economic Research, 1994.
63. Oldrich Vasicek. An Equilibrium Characterization of the Term Structure. *Journal of Financial Economics*, 5:177–188, 1977.
64. D. Wang. Pricing Defaultable Debt: Some Exact Results. *International Journal of Theoretical and Applied Finance*, 2(1):95–99, 1999.
65. David Gouming Wei and Dajiang Guo. Pricing Risky Debt: An Empirical Comparison of the Longstaff and Schwartz and Merton Models. *The Journal of Fixed Income*, pages 8–28, September 1997.
66. Chunsheng Zhou. A Jump-Diffusion Approach to Modeling Credit Risk and Valuing Securities. Working Paper, Federal Reserve Board, 1997.
67. Hao Zhou. Finite Sample Properties of Four Estimation Methods for a Square-Root Interest Rate Diffusion Model. *Journal of Computational Finance*, 2(5):89–122, Winter 2001/02.

Printing and Binding: Strauss GmbH, Mörlenbach

Lecture Notes in Economics and Mathematical Systems

For information about Vols. 1–454
please contact your bookseller or Springer-Verlag

Vol. 455: R. Caballero, F. Ruiz, R. E. Steuer (Eds.), Advances in Multiple Objective and Goal Programming. VIII, 391 pages. 1997.

Vol. 456: R. Conte, R. Hegselmann, P. Terna (Eds.), Simulating Social Phenomena. VIII, 536 pages. 1997.

Vol. 457: C. Hsu, Volume and the Nonlinear Dynamics of Stock Returns. VIII, 133 pages. 1998.

Vol. 458: K. Marti, P. Kall (Eds.), Stochastic Programming Methods and Technical Applications. X, 437 pages. 1998.

Vol. 459: H. K. Ryu, D. J. Slottje, Measuring Trends in U.S. Income Inequality. XI, 195 pages. 1998.

Vol. 460: B. Fleischmann, J. A. E. E. van Nunen, M. G. Speranza, P. Stähly, Advances in Distribution Logistic. XI, 535 pages. 1998.

Vol. 461: U. Schmidt, Axiomatic Utility Theory under Risk. XV, 201 pages. 1998.

Vol. 462: L. von Auer, Dynamic Preferences, Choice Mechanisms, and Welfare. XII, 226 pages. 1998.

Vol. 463: G. Abraham-Frois (Ed.), Non-Linear Dynamics and Endogenous Cycles. VI, 204 pages. 1998.

Vol. 464: A. Aulin, The Impact of Science on Economic Growth and its Cycles. IX, 204 pages. 1998.

Vol. 465: T. J. Stewart, R. C. van den Honert (Eds.), Trends in Multicriteria Decision Making. X, 448 pages. 1998.

Vol. 466: A. Sadrieh, The Alternating Double Auction Market. VII, 350 pages. 1998.

Vol. 467: H. Hennig-Schmidt, Bargaining in a Video Experiment. Determinants of Boundedly Rational Behavior. XII, 221 pages. 1999.

Vol. 468: A. Ziegler, A Game Theory Analysis of Options. XIV, 145 pages. 1999.

Vol. 469: M. P. Vogel, Environmental Kuznets Curves. XIII, 197 pages. 1999.

Vol. 470: M. Ammann, Pricing Derivative Credit Risk. XII, 228 pages. 1999.

Vol. 471: N. H. M. Wilson (Ed.), Computer-Aided Transit Scheduling. XI, 444 pages. 1999.

Vol. 472: J.-R. Tyran, Money Illusion and Strategic Complementarity as Causes of Monetary Non-Neutrality. X, 228 pages. 1999.

Vol. 473: S. Helber, Performance Analysis of Flow Lines with Non-Linear Flow of Material. IX, 280 pages. 1999.

Vol. 474: U. Schwalbe, The Core of Economies with Asymmetric Information. IX, 141 pages. 1999.

Vol. 475: L. Kaas, Dynamic Macroeconomics with Imperfect Competition. XI, 155 pages. 1999.

Vol. 476: R. Demel, Fiscal Policy, Public Debt and the Term Structure of Interest Rates. X, 279 pages. 1999.

Vol. 477: M. Théra, R. Tichatschke (Eds.), Ill-posed Variational Problems and Regularization Techniques. VIII, 274 pages. 1999.

Vol. 478: S. Hartmann, Project Scheduling under Limited Resources. XII, 221 pages. 1999.

Vol. 479: L. v. Thadden, Money, Inflation, and Capital Formation. IX, 192 pages. 1999.

Vol. 480: M. Grazia Speranza, P. Stähly (Eds.), New Trends in Distribution Logistics. X, 336 pages. 1999.

Vol. 481: V. H. Nguyen, J. J. Strodiot, P. Tossings (Eds.). Optimation. IX, 498 pages. 2000.

Vol. 482: W. B. Zhang, A Theory of International Trade. XI, 192 pages. 2000.

Vol. 483: M. Königstein, Equity, Efficiency and Evolutionary Stability in Bargaining Games with Joint Production. XII, 197 pages. 2000.

Vol. 484: D. D. Gatti, M. Gallegati, A. Kirman, Interaction and Market Structure. VI, 298 pages. 2000.

Vol. 485: A. Garnaev, Search Games and Other Applications of Game Theory. VIII, 145 pages. 2000.

Vol. 486: M. Neugart, Nonlinear Labor Market Dynamics. X, 175 pages. 2000.

Vol. 487: Y. Y. Haimes, R. E. Steuer (Eds.), Research and Practice in Multiple Criteria Decision Making. XVII, 553 pages. 2000.

Vol. 488: B. Schmolck, Ommitted Variable Tests and Dynamic Specification. X, 144 pages. 2000.

Vol. 489: T. Steger, Transitional Dynamics and Economic Growth in Developing Countries. VIII, 151 pages. 2000.

Vol. 490: S. Minner, Strategic Safety Stocks in Supply Chains. XI, 214 pages. 2000.

Vol. 491: M. Ehrgott, Multicriteria Optimization. VIII, 242 pages. 2000.

Vol. 492: T. Phan Huy, Constraint Propagation in Flexible Manufacturing. IX, 258 pages. 2000.

Vol. 493: J. Zhu, Modular Pricing of Options. X, 170 pages. 2000.

Vol. 494: D. Franzen, Design of Master Agreements for OTC Derivatives. VIII, 175 pages. 2001.

Vol. 495: I. Konnov, Combined Relaxation Methods for Variational Inequalities. XI, 181 pages. 2001.

Vol. 496: P. Weiß, Unemployment in Open Economies. XII, 226 pages. 2001.

Vol. 497: J. Inkmann, Conditional Moment Estimation of Nonlinear Equation Systems. VIII, 214 pages. 2001.

Vol. 498: M. Reutter, A Macroeconomic Model of West German Unemployment. X, 125 pages. 2001.

Vol. 499: A. Casajus, Focal Points in Framed Games. XI, 131 pages. 2001.

Vol. 500: F. Nardini, Technical Progress and Economic Growth. XVII, 191 pages. 2001.

Vol. 501: M. Fleischmann, Quantitative Models for Reverse Logistics. XI, 181 pages. 2001.

Vol. 502: N. Hadjisavvas, J. E. Martínez-Legaz, J.-P. Penot (Eds.), Generalized Convexity and Generalized Monotonicity. IX, 410 pages. 2001.

Vol. 503: A. Kirman, J.-B. Zimmermann (Eds.), Economics with Heterogenous Interacting Agents. VII, 343 pages. 2001.

Vol. 504: P.-Y. Moix (Ed.), The Measurement of Market Risk. XI, 272 pages. 2001.

Vol. 505: S. Voß, J. R. Daduna (Eds.), Computer-Aided Scheduling of Public Transport. XI, 466 pages. 2001.

Vol. 506: B. P. Kellerhals, Financial Pricing Models in Continuous Time and Kalman Filtering. XIV, 247 pages. 2001.

Vol. 507: M. Koksalan, S. Zionts, Multiple Criteria Decision Making in the New Millenium. XII, 481 pages. 2001.

Vol. 508: K. Neumann, C. Schwindt, J. Zimmermann, Project Scheduling with Time Windows and Scarce Resources. XI, 335 pages. 2002.

Vol. 509: D. Hornung, Investment, R&D, and Long-Run Growth. XVI, 194 pages. 2002.

Vol. 510: A. S. Tangian, Constructing and Applying Objective Functions. XII, 582 pages. 2002.

Vol. 511: M. Külpmann, Stock Market Overreaction and Fundamental Valuation. IX, 198 pages. 2002.

Vol. 512: W.-B. Zhang, An Economic Theory of Cities.XI, 220 pages. 2002.

Vol. 513: K. Marti, Stochastic Optimization Techniques. VIII, 364 pages. 2002.

Vol. 514: S. Wang, Y. Xia, Portfolio and Asset Pricing. XII, 200 pages. 2002.

Vol. 515: G. Heisig, Planning Stability in Material Requirements Planning System. XII, 264 pages. 2002.

Vol. 516: B. Schmid, Pricing Credit Linked Financial Instruments. X, 246 pages. 2002.

Vol. 517: H. I. Meinhardt, Cooperative Decision Making in Common Pool Situations. VIII, 205 pages. 2002.

Vol. 518: S. Napel, Bilateral Bargaining. VIII, 188 pages. 2002.

Vol. 519: A. Klose, G. Speranza, L. N. Van Wassenhove (Eds.), Quantitative Approaches to Distribution Logistics and Supply Chain Management. XIII, 421 pages. 2002.

Vol. 520: B. Glaser, Efficiency versus Sustainability in Dynamic Decision Making. IX, 252 pages. 2002.

Vol. 521: R. Cowan, N. Jonard (Eds.), Heterogenous Agents, Interactions and Economic Performance. XIV, 339 pages. 2003.

Vol. 522: C. Neff, Corporate Finance, Innovation, and Strategic Competition. IX, 218 pages. 2003.

Vol. 523: W.-B. Zhang, A Theory of Interregional Dynamics. XI, 231 pages. 2003.

Vol. 524: M. Frölich, Programme Evaluation and Treatment Choise. VIII, 191 pages. 2003.

Vol. 525:S. Spinler, Capacity Reservation for Capital-Intensive Technologies. XVI, 139 pages. 2003.

Vol. 526: C. F. Daganzo, A Theory of Supply Chains. VIII, 123 pages. 2003.

Vol. 527: C. E. Metz, Information Dissemination in Currency Crises. XI, 231 pages. 2003.

Vol. 528: R. Stolletz, Performance Analysis and Optimization of Inbound Call Centers. X, 219 pages. 2003.

Vol. 529: W. Krabs, S. W. Pickl, Analysis, Controllability and Optimization of Time-Discrete Systems and Dynamical Games. XII, 187 pages. 2003.

Vol. 530: R. Wapler, Unemployment, Market Structure and Growth. XXVII, 207 pages. 2003.

Vol. 531: M. Gallegati, A. Kirman, M. Marsili (Eds.), The Complex Dynamics of Economic Interaction. XV, 402 pages, 2004.

Vol. 532: K. Marti, Y. Ermoliev, G. Pflug (Eds.), Dynamic Stochastic Optimization. VIII, 336 pages. 2004.

Vol. 533: G. Dudek, Collaborative Planning in Supply Chains. X, 234 pages. 2004.

Vol. 534: M. Runkel, Environmental and Resource Policy for Consumer Durables. X, 197 pages. 2004.

Vol. 535: X. Gandibleux, M. Sevaux, K. Sörensen, V. T'kindt (Eds.), Metaheuristics for Multiobjective Optimisation. IX, 249 pages. 2004.

Vol. 536: R. Brüggemann, Model Reduction Methods for Vector Autoregressive Processes. X, 218 pages. 2004.

Vol. 537: A. Esser, Pricing in (In)Complete Markets. XI, 122 pages, 2004.

Vol. 538: S. Kokot, The Econometrics of Sequential Trade Models. XI, 193 pages. 2004.

Vol. 539: N. Hautsch, Modelling Irregularly Spaced Financial Data. XII, 291 pages. 2004.

Vol. 540: H. Kraft, Optimal Portfolios with Stochastic Interest Rates and Defaultable Assets. X, 173 pages. 2004.

Vol. 541: G.-Y. Chen, Vector Optimization (planned).

Vol. 542: J. Lingens, Union Wage Bargaining and Economic Growth. XIII, 199 pages. 2004.

Vol. 543: C. Benkert, Default Risk in Bond and Credit Derivatives Markets. IX, 135 pages. 2004.

Vol. 545: R. Hafner, Stochastic Implied Volatility. XI, 229 pages. 2004.

Vol. 546: D. Quadt, Lot-Sizing and Scheduling for Flexible Flow Lines. XVIII, 227 pages. 2004.